PRINCIPLES AND PROBLEMS OF
BIBLICAL TRANSLATION

PRINCIPLES AND PROBLEMS OF BIBLICAL TRANSLATION

SOME REFORMATION CONTROVERSIES AND THEIR BACKGROUND

BY

W. SCHWARZ

LECTURER IN GERMAN, UNIVERSITY
COLLEGE, LONDON

CAMBRIDGE
AT THE UNIVERSITY PRESS
1955

PUBLISHED BY
THE SYNDICS OF THE CAMBRIDGE UNIVERSITY PRESS

London Office: Bentley House, N.W. 1
American Branch: New York
Agents for Canada, India, and Pakistan: Macmillan

Printed in Great Britain at the University Press, Cambridge
(Brooke Crutchley, University Printer)

CONTENTS

Foreword by DR C. H. DODD		page vii
Preface		ix
List of abbreviations		xiii
I	THE BIBLE AND THE TRANSLATOR	1
II	DISCUSSIONS ON THE ORIGIN OF THE SEPTUAGINT	17
III	THE TRADITIONAL VIEW	45
IV	THE PHILOLOGICAL VIEW: REUCHLIN	61
V	THE PHILOLOGICAL VIEW: ERASMUS OF ROTTERDAM	92
VI	THE INSPIRATIONAL VIEW: LUTHER	167
Select Bibliography		213
Bible Index		215
General Index		217

FOREWORD

The art of translation is one in which complete success is for ever impossible. No translation is final, even if it be so brilliantly executed as to attain the status of a classic in its own language. The spate of new biblical translations in recent years testifies to a certain uneasiness about the finality of even our most highly respected versions, whether of the school of Douai or of King James's men, or of Martin Luther—for the movement is by no means confined to the 'English speaking world'. Each new version is criticized in turn, and often the criticisms turn upon presuppositions which the critic has hardly made clear to himself.

Those who are hardy enough to put their hands to the fascinating but elusive task of Bible translation soon become aware of a host of problems which must be settled one way or the other in the course of the work. Are the actual words of *das heilige Original* so full of significance that they must be rendered, as far as possible, word for word? Or must the subject-matter overrule the words, so that the rendering represents the meaning of the original as the translator understands it, and as he would himself express it in addressing his own contemporaries and countrymen? In doing so, does he act mainly as scholar, or stylist—or prophet? And what if in pursuing his aim he should break so violently with tradition as to obscure terms of high doctrinal significance? Should the translator, then, be guided, or even overruled, by the theologian? Or is this an opportunity of confronting the theologian afresh with the plain sense of Holy Scripture which, as he himself admits, is the permanent standard of reference for all Christian doctrine?

Such questions cannot be either evaded or answered out of hand. That they are no new questions we may learn from this book. They were all raised, and all discussed amply, acutely, and often with deep appreciation of the principles involved, in the period of the Reformation. But neither were they new questions

FOREWORD

then. The controversy between Erasmus and Luther was essentially the same as the controversy between Jerome and Augustine centuries earlier; and indeed the main issues may be said to have been raised as soon as the Septuagint made a bid for acceptance as the equivalent of the Hebrew Scriptures. All this we may read, set forth with learning and penetration, in Dr Schwarz's book. A study of it will be illuminating (*experto crede*) to the translator, salutary to the critic, and profitable to anyone interested either in the art of translation in general or, particularly, in the enterprise of making the biblical writings accessible in the vernacular to the mind of this generation.

C. H. DODD

PREFACE

It is now a commonplace for students of the Bible to pay attention to the literary value of Holy Writ. They do not, of course, wish to neglect its religious content but they maintain that the literary and religious values are interconnected.

Yet this has by no means always been the case. It was in the sixteenth century, the age of Erasmus and humanism, of Luther and the Reformation, that this decisive change took place in men's attitude towards the Bible. The origins, course and results of this revolution in thought constitute the central theme unifying the studies contained in this book. The results of humanistic theory have been handed down to our own age and are a basis of modern biblical exegesis. Our very dependence on the thought of the humanists, however, makes it difficult to view their work in historical perspective and to appreciate the thought which prevailed before their time. Their study of classical literature led them to the belief that no translation of a literary work could replace, or be completely equivalent to, its original. This precept when transferred to Holy Scripture implied that the Bible must be interpreted from the Hebrew and Greek texts and not from the Latin translation, the Vulgate. When it was first advocated this was indeed a revolutionary idea. It was new and yet linked with tradition, for Latin Fathers of the Church, especially St Jerome, had postulated it for the Bible many centuries earlier.

St Jerome's work had been known throughout the Middle Ages. Why then was the demand of the humanists something new? The Bible used in Western Europe during the Middle Ages was the Vulgate. It was the version used in church. Its wording was the basis of exegesis and of the translation into the vernacular languages. Moreover, the scholastic philosophy, closely allied or even subordinated to theology, relied on the text of the Vulgate. A combination of theology and philosophy protected the authority of the Vulgate. The humanists had to fight these forces

PREFACE

throughout their lives. There was an additional reason which helped the defenders of the Vulgate. During the Middle Ages the value of the Bible in the original languages had been forgotten. Indeed, only at some short periods during the Middle Ages had anyone sufficient knowledge of Hebrew and Greek to study them. For a long time nobody had compared the Vulgate with the original texts and nobody knew, therefore, whether there were any differences between these versions. Reuchlin and Erasmus were the first to publish comparisons of the Latin translation with the originals. They found the Vulgate wanting. They could show that ambiguities in the Vulgate which had given rise to long commentaries and even philosophical speculations could be explained from the wordings of the original languages. This discovery dealt a deadly blow to medieval exegesis and scholasticism. No rational answer could be found to it. It could be said that the authoritative position of the Vulgate was imperilled and with it the doctrines of the Church which were based on the Latin text. The humanists, however, could maintain that they did not endanger the Church but intended only to reconstitute the Latin wording of the Vulgate and to contribute to a better understanding of God's word.

The humanists had found a new method of Bible interpretation, freed from scholastic and theological speculations. They believed they were able to comprehend God's word by studying the Hebrew and Greek texts. Their interpretations and their translations were the outcome of their life's work. It is therefore indispensable to discuss certain biographical details of men like Reuchlin and Erasmus, the inaugurators of modern exegesis, and to relate their thought to the then prevailing scholastic philosophy.

The controversy between the Schoolmen and the humanists grew in intensity and bitterness when Luther came to the fore. There were many points of agreement between the humanists and the men of the Reformation. Therefore they wished to avoid open conflict between themselves. Yet the differences separating Erasmus and Luther were such as to lead to open dispute. From the very beginning Erasmus was convinced that learning could

PREFACE

result in knowledge of the past and the creation of a new civilization. Luther, however, could not accept this humanistic belief.

When he read Erasmus's commentary on the New Testament, he understood that the gulf separating himself from Erasmus could not easily be bridged. Before going into this argument it must be understood that their controversy was not entirely new. It was a controversy that had occurred centuries before between St Jerome and St Augustine, who had clearly described the different methods with which to approach Bible interpretation and translation. And even St Jerome and St Augustine were indebted to earlier thought. Indeed, it is possible to trace the nature of the controversy between Erasmus and Luther back to the first century A.D.

As the title suggests, this book deals with the principles governing biblical translation. Translations as such will only be used to elucidate the principles developed by the translator. The question of the correspondence between principle and practice will not be raised.

Some details of philological research have been omitted in the chapter on Luther. They are important since they support my attitude to Luther. A paper on these points is being published in *The Journal of Theological Studies*, 1955.

I wish to mention some technical points which may be of interest to the reader. References to secondary literature dealing with biblical translations have been reduced to the barest minimum. I have refrained from engaging in polemics and have merely discussed what I believe to be the result of the interpretation of the original sources. I hope, however, that I have mentioned all those authors whose views I found relevant to the subject of this book. The Select Bibliography combined with the references in the footnotes should be of use to the reader.

All the abbreviations found in early printed books have been expanded, the spelling of 'u' and 'v' has generally been regulated in accordance with modern usage, 'ę' has been replaced by 'ae' but otherwise the original spelling has been retained.

PREFACE

I have always referred to the pages of Migne, *Patrologia Graeca* and *Patrologia Latina*, since they are easily accessible, although modern editions have been used throughout.

It is a pleasant duty to thank all those who have helped me with their encouragement, advice and criticism, and who after reading the manuscript have drawn my attention to many details of substance and form which needed further thought. It is impossible to mention all their names but I would like to refer especially to Professor Norman H. Baynes, Miss Barbara Flower, Professor L. W. Forster, Dr J. F. Lockwood, Master of Birkbeck College, Professor C. P. Magill, Mr H. D. Sacker, Mr J. M. L. Trim, Dr Elizabeth M. Wilkinson, Professor L. A. Willoughby, and my wife, whose constructive criticism and co-operation have been invaluable to me. I am grateful to the Syndics of the Cambridge University Press for accepting my book for publication and to the staff of the Press for producing it so carefully. I am greatly obliged to the staffs of the Reading Room of the British Museum, of Dr Williams's Library, and of the Library of University College, London, for their unfailing help and courtesy. I would like to thank especially Professor C. H. Dodd for the interest he has taken in the book, the support he has given it throughout, and especially for writing the Foreword.

W. S.

LONDON
September 1954

LIST OF ABBREVIATIONS

A.D.B. Allgemeine Deutsche Biographie.

A.H.A. Abhandlungen der Heidelberger Akademie der Wissenschaften Philosophisch-historische Klasse.

Apologia, N.T.1, N.T.2. Apologia, as found in Erasmus's first and second edition of the New Testament.

B.D.G. K. Schottenloher, *Bibliographie zur Deutschen Geschichte im Zeitalter der Glaubensspaltung 1517–1585*, vols. I–VI (Leipzig, 1933–40).

B.Er.1. Bibliotheca Erasmiana. Listes sommaires, ser. 1–3 (Ghent, 1893).

B.Er.2. Bibliotheca Erasmiana; Bibliographie des œuvres d'Erasme (Ghent, 1897–1908).

C.R. Corpus Reformatorum.

Contra Morosos. Capita Argumentorum contra Morosos Quosdam et Indoctos, as found in Erasmus's New Testament of 1519.

E.A. Luther, *Werke*, vols. I–LXVII (Frankfurt and Erlangen, 1826–57).

E.E. Opus Epistolarum Des. Erasmi Roterodami, edd. P. S. and H. M. Allen, and H. W. Garrod, vols. I–XI (Oxford, 1906–47).

E.E.... = N.: e.g. *E.E.* vol. I, no. 108 (74–6) = *N.* vol. I, p. 221. This abbreviation refers to Erasmus's letter in P. S. Allen's edition, vol. I, no. 108, ll. 74–6. The translation is taken from: *The Epistles of Erasmus from his earliest Letters to his fifty-first Year arranged in order of time. English Translations*...by F. M. Nichols, vols. I–III (London, 1901–18).

Hauck, *Realencyclopaedie.* Realencyklopaedie für protestantische Theologie und Kirche, 3rd ed., ed. D. A. Hauck, vols. I–XXIV (Leipzig, 1896–1913).

I.Gl. Interlinear gloss.

L.B. Desiderii Erasmi Roterodami Opera Omnia, 10 vols. (Leyden, 1703–6).

L.E. Luther, *Briefwechsel*, 8 vols. Weimar Ausgabe (*W.A.*).

Me.E. Melanchthon's *Epistolae* in *C.R.* vol. I (Halle, 1834).

M.Gl. Marginal Gloss.

Mo.E. The correspondence of Sir Thomas More, ed. E. F. Rogers (Princeton, 1947).

LIST OF ABBREVIATIONS

Mu.E. Der Briefwechsel des Conradus Mutianus, ed. K. Gilbert (Geschichtsquellen der Provinz Sachsen und angrenzender Gebiete, hrsggb. von der historischen Commission der Provinz Sachsen, vol. 18 (Halle, 1890)).

N. F. M. Nichols, see *E.E.*... = N.

N.T.1. Erasmus, New Testament, edition of 1516.

N.T.2. Erasmus, New Testament, edition of 1519.

Paraclesis, N.T.1, N.T.2. Paraclesis as found in the first and second editions of Erasmus's New Testament.

P.G. Migne, J. P., Patrologiae Cursus Completus... Scriptores Ecclesiae Graecae.

P.L. Migne, J. P., Patrologiae Cursus Completus... Scriptores Ecclesiae Latinae.

Ratio. Ratio seu Compendium Verae Theologiae. (All the quotations refer to the edition found in Erasmus's New Testament of 1519.)

R.E. Johann Reuchlins Briefwechsel gesammelt und herausgegeben von L. Geiger, Bibliothek des Litterarischen Vereins in Stuttgart, vol. CXXVI (Tübingen, 1875).

S. Scholia.

S.B.A. Sitzungsberichte der Akademie der Wissenschaften, Berlin, Philos.-histor. Klasse.

Scheel. O. Scheel, 'Dokumente zu Luthers Entwicklung (bis 1519)', Sammlung ausgewaehlter Kirchen- und Dogmengeschichtlichen Schriften. N.F. No. 2, 2nd ed. (Tübingen, 1929).

S.H.A. Sitzungsberichte der Heidelberger Akademie der Wissenschaften, Philos.-histor. Klasse.

S.W.A. Sitzungsberichte der Akademie der Wissenschaften, Wien, Philos.-histor. Klasse.

T.R. Luther, Tischreden, vols. I–VI, Weimar Ausgabe (*W.A.*).

W.A. Luther, Werke, vols. I–LVII, Weimar Ausgabe (Weimar, 1883–   ).

W.A.B. Luther, Die Deutsche Bibel, vols. I–IX, Weimar Ausgabe (*W.A.*).

CHAPTER I

THE BIBLE AND THE TRANSLATOR

Every modern translation is an interpretation of the original work. The modern translator attempts to reproduce in his own language the thought contained in the work as a whole, in each sentence and even each word within its context. His task is not finished here, however. He must strive to convey the artistic form of the original, preserving the author's use of imagery, the rhythm of his sentences, and in the case of a poem its metre and rhythm. He tries to recreate in his rendering the very tone and atmosphere of the foreign literary work. The modern translator must therefore be able to interpret a given text in every aspect and to reproduce this interpretation in his own language. These requirements cause his work to be short-lived, for the interpretation is dependent on the currents of thought prevalent in his own time. Changes in thought, advances in scholarship, not to speak of the development of the living language, are bound to make earlier research and earlier translations obsolete.

Bible translations seem to defy this rule. The Greek translation of the Old Testament was made in the third century B.C. and is still in use. The Latin translation of Scripture goes back to about A.D. 400. The English Authorized Version was published in 1611. Yet not all Bible translations have shown the longevity of these versions. Others have deliberately appealed more directly to the tastes and opinions of their own time. Their authors have considered the Bible as a work of literature and have applied to it those principles of literary translation evolved by the humanists—above all Erasmus of Rotterdam—and their successors. These principles are clearly stated in the title of Edward Harwood's version of the Bible of 1768: *A Liberal Translation of the New Testament; being an Attempt to translate the Sacred Writings with the same Freedom, Spirit, and Elegance, With which other English*

Translations from the Greek Classics have lately been executed. This translation is based on a view of New Testament Greek which has been superseded. In the eighteenth century the New Testament was considered to be part of Greek classical literature, whilst in the twentieth it is placed together with the language of the papyri. Hence Mr J. Moffatt has endeavoured 'to translate the New Testament exactly as one would render any piece of contemporary Hellenistic prose, hoping to convey to the reader something of the direct homely impression made by the original upon those for whom it was written'.[1] In both cases the guiding principle is that the New Testament as a literary work must be translated like any other literary work of the same period. The special character and atmosphere of the original can be conveyed to the reader by reproducing the linguistic character.

That 'the immortal theme' should be transmitted in a language which is both intelligible and impressive is indeed a basic assumption of modern translation. It has guided the translators of the American Revised Standard Version of 1946 who say that their aim was to make 'a translation which is native to the forms of speech of our present world.... Let it be said for the makers of this translation that they have tried to make it a sensitive transmission of the immortal themes of the New Testament to this generation's mind and heart and ear.'[2] It has governed those who made the report for a proposed new translation into English when they say:

1. The work shall be a new translation, not a revision of the Authorised Version or of the Revised Version, having as its object to render the original into contemporary English and avoiding all archaic words and forms of expression; the second personal pronoun singular shall be employed only in prayer.

2. Regard shall be paid to the native idiom and current usage of the English language, and Hebraisms, Grecisms, and other un-English expressions shall be avoided; freedom shall be employed in altering the

---

[1] Preface to the edition of 1935, p. xliv.
[2] W. R. Bowie, 'The Use of the New Testament in Worship', in Luther A. Weigle, *An Introduction to the Revised Standard Version of the New Testament* (International Council of Religious Education, 1946), p. 63.

construction of the original when that is considered necessary in order to make the meaning intelligible in English; and the advice of one or more literary men shall be sought regarding the language of the translation.[1]

The same feeling has led the Archbishop of Westminster to write in the Preface to Mgr Knox's version of the New Testament: 'There is an engaging freshness about the narrative which captivates the reader, and a clarity of expression which removes the obscurity of not a few passages in former translations.' Mgr Knox clearly states that a translator of the Bible should be governed by H. Belloc's precept for the translation of secular literature. He should ask 'What would an Englishman have said to express this?' Therefore he thinks it necessary to 'transmute' the original sentences into English idiomatic language.[2] This demand makes it necessary to re-adapt the existing version to the changes of the living language. It is interesting to note that John Worsley, who translated the New Testament in 1770, drew attention to this fact when in the 'Author's Advertisement' he wrote:

The English Translation of the BIBLE in the Reign of King James I is, no doubt, a very good one; and justly so esteemed to this day; though it be above a hundred and fifty years old; but it is not to be wondered at if some words and phrases, then in use and well understood, should by this time become obsolete and almost unintelligible to common readers.... The principal attempt therefore of this Translation is... to make the form of expression more suitable to our *present* language. For as the English tongue, like other living languages, is continually changing, it were to be wished that the translation of the sacred oracles could be revised by public authority, and reduced to *present* forms of writing and speaking, at least once in a century: but though this be not allowed for *public* use, it is to be hoped some *private* persons may receive benefit by that which is now offered.[3]

Worsley makes a difference between a version of the Bible for private reading and one used for services. Indeed, the only

---

[1] *The Church of Scotland. Report of Special Committee anent proposed New Translation of the Bible* (May 1947), pp. 479–80.
[2] R. A. Knox, *On Englishing the Bible* (London, 1949), *passim*; see esp. pp. 4 and 19.
[3] London, 1770, fol. A3ᵛ.

versions not subject to continuous revision and replacement are those officially recognized and used in Church. As a result the authoritative versions, be they the original texts or translations, will at some time be criticized for their archaic language. It is therefore not completely justified to say: 'The English have always been accustomed to having an archaic Bible.'[1] Even the Vulgate was criticized for shortcomings in its style by some sixteenth-century humanists such as Reuchlin and Erasmus of Rotterdam.

The reasons for the longevity of authoritative versions are clear. It will at once be seen that the text of the Bible and the wordings of the prayers derived from it cannot be changed to comply with every 'generation's mind and heart and ear'. Every religious community must insist on tradition and the permanence of the wording of the holy text and its prayers. The living language changes but the stability of religious belief demands the stability of the wording of Holy Writ. The knowledge that our fathers and forefathers have said the same words as our own generation, memories of having learned these words in early childhood, these and more sentimental feelings are reasons for esteeming the official version of the Bible and the prayers, even if the archaic language causes single words or complete passages to be no longer fully understood. The solemnity connected with the old-fashioned language may even enhance its value within the community. The authoritative Bible text is considered as being of venerable sacredness, and it is therefore difficult to change it. St Augustine reported disturbances in an African community directed against the use of a new version of the Latin Bible. In more modern times the same feeling of reverence has been extended towards the English Authorized Version even when its language was severely criticized. Edward Harwood, whose translation of 1768 has been mentioned above, wrote:

The author knew it to be an arduous and invidious attempt to make the phrase of these celebrated writers [Hume, Robertson etc.] the vehicle

[1] R. A. Knox, *op. cit.* p. 13.

of inspired truths, and to diffuse over the sacred page the elegance of modern English, conscious that the bald and barbarous language of the old vulgar version hath acquired a venerable sacredness from length of time and custom, and that every innovation of this capital nature would be generally stigmatized as the last and most daring enormity.[1]

In 1881 a reviewer of the Revised Version wrote in a similar vein in the *Edinburgh Review*:

So long a period of time has elapsed since the last revision of the New Testament, and so great is its superiority to all the preceding translations, that we are exposed to the strong temptation of attaching to the so-called Authorised version an attribute of finality to which those who were concerned in its production, laid no claim....[2]

This esteem for an existing Bible version, the feeling of reverence towards it must be taken into account by anyone making a translation which he would wish to be recognized by the community. It is, I think, true to say that the desire not to alter the existing version of the Bible unnecessarily, has been a guiding principle of English Bible translation, at least since the Reformation. There has, until recently, been no intention of making a new translation. The Authorized Version expresses this regard for earlier versions in the Preface to the Reader:

Truly (good Christian Reader) wee neuer thought from the beginning, that we should neede to make a new Translation, nor yet to make of a bad one a good one,...but to make a good one better, or out of many good ones, one principall good one, not iustly to be excepted against; that hath bene our indeauour, that our marke.[3]

This is in full harmony with 'The Rules to be observed in the Translation of the Bible' which were given to the translators of the Authorized Version. The first of these 'Rules' reads:

The ordinary Bible read in the Church, commonly called the *Bishops' Bible*, to be followed, and as little altered as the truth of the original will permit.[4]

---

[1] London, 1768, p. v. The interpretation of this preface is not in agreement with C. S. Lewis, *The Literary Impact of the Authorised Version* (London, 1950), p. 20.
[2] July 1881, p. 158.   [3] Edition of 1611, fol. B^v.
[4] Quoted by J. Isaacs, 'The Authorized Version and After', in H. Wheeler Robinson, *The Bible in its Ancient and English Versions* (Oxford, 1940), p. 199.

If the instructions given for the making of the Revised Version in 1870 are compared with these passages, their similarity is startling. They read:

To introduce as few alterations as possible into the Text of the Authorized Version, consistently with faithfulness.[1]

Those who had to make a new version or, it would be better to say, the revision, were bound to retain the old text wherever possible. The history of the English official Bible is indeed a history of revisions.

The same tendency can be observed in America: there in 1901 the American Standard Revision is a revision of the English Authorized Version, whilst the edition of 1946, called the Revised Standard Version, is a revision of the 1901 revision.

Underlying the opposition to changes of an authoritative version is not only the wish to preserve the tradition of the religious books of the community, but also the feeling that their texts are final and sacred. Hallowed by age-long use they are considered to be God's word, just as the original words which were given by God to man. The authoritative translation is thought by the lay community and by the priesthood to be a faithful, final, and sacred reproduction of Holy Writ which must not be changed. But, it may be asked, how can a translation, made by man with his shortcomings, be equal to the perfection of Scripture? Is man allowed to transfer God's message, given in a certain language, into another language by replacing His wording with another text of the same meaning? Since every translation is an interpretation, how can the interpretation of a human being, however conscientiously it may have been executed, be a substitute for God's word? Does not the translation reflect the mind of man and in this way belittle God's revelation? Does it not take away at least part of the depth of meaning that is implied in the original? Can the translation into

---

[1] C. J. Cadoux, 'The Revised Version and After', in Wheeler Robinson, *op. cit.* p. 242.

another language reproduce all the thought of the revelation, especially as no human being can fully understand its mystery?

The divine origin of Scripture gives the text a special significance. Every word, even the single letters, contains God's mystery. The order of words may have a meaning hidden from the translator but perhaps to be revealed in the future. The translator who sees these questions, as indeed they were seen at an early stage, must needs despair in his task. How should he, within his human limitations, be able to understand even part of God's word and how could he hope to render it without destroying the mystery contained in it? He is free to choose between the possible interpretations of ambiguous sentences and to decide on a special wording which against his wish may change the shade of meaning. This freedom, which allows him to expound God's word after his own limited fashion, must needs terrify him. One way out of this dilemma is to follow an authority which furnishes him with a reliable exegesis of at least those parts which are of the greatest religious significance. Such an authority, whilst imposing its will and thought on him, enables him to rest assured that his work, in spite of shortcomings, is correct. It goes without saying that he must believe in this authority and follow its exegesis without hesitation or doubt. Protected by it, he may feel free to do his work. Such an authority is the Church to which the translator belongs. The tenets and beliefs of this community will help him in his task.

Authority, however, will be of support to the translator only if he willingly accepts its rules. Otherwise he will only feel the chains imposed on his thought. This dual role which authority is bound to play has been most clearly characterized in an Encyclical Letter of Pope Pius XII in the following words:

Authority, by their way of it, is a drag on progress, is a bar to the development of science; there are some non-Catholics who think of it as a bridle, which forcibly restrains some few enlightened theologians from revolutionizing the whole system they teach. Is not this Authority [of the Church] a sacred trust, an exact and all-embracing standard of measurement which every theologian must use? Has not our Lord Christ committed to it the task of guarding, preserving, interpreting the whole deposit of faith, not only Sacred Scripture, but the tradition which is no less divine in origin?[1]

The religious authorities jealously guard their own tradition, which is an essential part of their religious belief. From their point of view they have the right to supervise the translator's work in every detail, to condemn it if found faulty, to ask for corrections, or to approve of it, and to declare it a faithful and even an authentic version. This claim of the authorities is based on the knowledge or the assumption that the traditional exegesis of a sacred text is bound up with the belief of the community and that therefore the community is conjoined with the exposition as handed down from their forefathers. The traditional exegesis is justified by its long continuity which may claim to go back to the founder or to those who were able to recall his teaching or both. It may have been written down for all posterity by men who, by the grace of God, were inspired and whose words are valid and final. All those who belong to a certain religious group will follow these tenets and the interpretations derived from or leading to them, whilst those who do not share the belief in the inspiration of these expounders will reject the tradition based on them. For this reason different religious communities, while revering the same sacred texts, follow their own individual interpretations and have their own different translations: the Roman Catholic version of the Bible must, of necessity, differ from that of the other Christian communities in all those parts in which there are doctrinal disagreements between the different Churches.

Religious belief is therefore the corner-stone of biblical interpretation. No religious community can admit that any external authority has any say in the exegesis of its holy books. This claim has been most vigorously enforced in the Roman Catholic Church.

From a very early stage in its history the Catholic Church has

[1] Encyclical Letter *Humani Generis* of 12 August 1950, translated by Mgr R. A. Knox (Catholic Truth Society, London, 1950), p. 10.

claimed to be the sole authority entitled to give the correct doctrinal exegesis of Scripture. The apostles, it is maintained, transmitted Holy Scripture to their successors, that is to the Church, which thus has the duty and the right to expound the meaning of God's word. In this the Church cannot err. The Bible and the Roman Catholic Church are therefore an indissoluble unity. These arguments lead to the following conclusions: whenever the Church has authoritatively expounded any verse of the Bible, this interpretation is of necessity correct and final and no member of the Church can legitimately disagree with it. This tradition of the Church is clearly stated in the Encyclical Letter *Providentissimus Deus* given by Pope Leo XIII on 18 November 1893. Referring to the Fathers of the Church in whose works many interpretations of Bible verses are found, it maintains that the Holy Fathers 'are of supreme authority, whenever they all interpret in one and the same manner any text of the Bible, as pertaining to the doctrine of faith or morals; for their unanimity clearly evinces that such interpretation has come down from the Apostles as a matter of Catholic faith'. But even if they speak 'in their capacity as doctors, unofficially' their opinion is 'of very great weight', not only because they possess great knowledge but because on them 'God has bestowed a more ample measure of His light'. Yet '"in those things which do not come under the obligation of faith, the Saints were at liberty to hold divergent opinions just as we ourselves are", according to the saying of St Thomas'.[1] This last sentence seems to open a way to individual free reasoning whenever the Fathers do not agree. This freedom is emphasized in the Encyclical Letter *Divino Afflante Spiritu*, given by Pius XII on 30 September 1943 in these words:

Let them remember above all that the rules and laws laid down by the Church are concerned with the doctrine of faith and morals; and that

---

[1] All the quotations of the Encyclical Letter of Leo XIII are taken from *The Holy Bible translated from the Latin Vulgate and diligently compared with other editions in divers languages (Douay, A.D. 1609, Rheims, A.D. 1582) published as revised and annotated by authority. With a Preface by the Cardinal Archbishop of Westminster*. The above passages are pp. xxiii and xxix.

among the many matters set forth in the legal, historical, sapiential, and prophetical books of the Bible there are only a few whose sense has been declared by the authority of the Church, and that there are equally few concerning which the opinion of the holy Fathers is unanimous. There consequently remain many matters, and important matters, in the exposition and explanation of which the sagacity and ingenuity of Catholic interpreters can and ought to be freely exercised, so that each in the measure of his powers may contribute to the utility of all, to the constant advancement of sacred learning, and to the defence and honour of the Church.[1]

But when the Church has authoritatively spoken, no further discussion is allowed:

In things of faith and morals, belonging to the building up of Christian doctrine, that is to be considered the true sense of Holy Scripture which has been held and is held by our Holy Mother the Church whose place it is to judge of the true sense and interpretation of the Scriptures; and therefore that it is permitted to no one to interpret Holy Scripture against such sense or also against the unanimous agreement of the Fathers.[2]

In this connexion some questions may be raised which show the limits imposed upon the translator by the religious authority.

The history of European Bible translation teaches the student to appreciate the importance of the existence of an authoritative version such as the Latin Vulgate which was declared the authentic text only very late, in a decree of the Council of Trent, dated 8 April 1546.[3] This decree which legalized the status of the Vulgate states that 'the Vulgate approved through long usage during so many centuries be held authentic in public lectures, disputations, preaching and exposition, and that nobody dare or presume to reject it under any pretext'. The meaning of 'authentic' version in

[1] Translated by Canon C. D. Smith (Catholic Truth Society, 1944), p. 30.
[2] Encyclical Letter of 1893, *loc. cit.* p. xxi. This is a quotation from the Council of the Vatican.
[3] The text of the decree of 1546 is found in *Canones et Decreta Concilii Tridentini*, ed. F. Schulte–A. L. Richter (Leipzig, 1853), p. 12. For its origin and interpretation see P. Richard, *Concile de Trente*, in C. H. Hefele, *Histoire des Conciles d'après les documents originaux*, vol. IX, 1 (Paris, 1930), and H. Höpfl, 'Beiträge zur Geschichte der Sixto-Klementinischen Vulgata', *Biblische Studien*, vol. XVIII, 1–3 (Freiburg, 1913), pp. 1 ff. and esp. pp. 22 ff.

this sentence is that such a version replaces the original in every respect. The authentic version and not the original is therefore the basis for the interpretation and translation of the Bible.

The final renewal of this decree was made in the Encyclical Letter of 1893 mentioned above, which says that when instructing students in Scripture

the Professor, following the tradition of antiquity, will make use of the Vulgate as his text; for the Council of Trent has decreed that 'in public lectures, disputations, preaching and exposition', the Vulgate is the 'authentic' version; and this is the existing custom of the Church. At the same time the other versions which Christian antiquity has approved, should not be neglected, more especially the more ancient MSS. For although the meaning of the Hebrew and the Greek is substantially rendered by the Vulgate, nevertheless wherever there may be ambiguity or want of clearness, the 'examination of older tongues' to quote St Augustine, will be useful and advantageous.[1]

It is therefore clear that a Roman Catholic translation of the Bible must be based on the Vulgate, although account may be taken of the readings in Hebrew and Greek. Throughout the Middle Ages all the renderings of Holy Scripture were made from the Latin text of the Vulgate. Generally speaking, Hebrew and Greek were not known before the sixteenth century and only then was the importance of these languages for the study of the Bible recognized. The Roman Catholic Church has retained the Vulgate as its 'authentic' version which is the basis of every translation and interpretation.

The translator of the Vulgate who does not neglect the original texts may easily face difficulties whenever there is a disagreement between the wording in the different languages. He may even consider that the text of the Vulgate is in need of corrections. The precepts of Leo XIII's Encyclical Letter are flexible enough to leave him perturbed, for they read:

It is true, no doubt, that copyists have made mistakes in the text of the Bible; this question, when it arises, should be carefully considered on its merits, and the fact not too easily admitted, but only in those passages where the proof is clear.[2]

---

[1] *Loc. cit.* pp. xix–xx.   [2] *Loc. cit.* p. xxx.

THE BIBLE AND THE TRANSLATOR

Should the translator change the text of the Vulgate if he is convinced that the text is corrupt, or would it not be wiser to follow the approved wording? This question has been asked by Mgr Knox, who prefers to solve it in favour of the traditional text. In his view the translator should not alter his original. This decision shows the great importance that must be attached to the existence of an authoritative version. Mgr Knox's words are:

I have even denied myself the privilege claimed by the latest American revisers, of going back behind the Clementine edition, and taking the Vulgate as it stands (say) in Wordsworth and White's collation of it. The American version, for example, in Acts xvii. 6, has 'these men who are setting the world in uproar'. That is quite certainly the true reading; but a bad copyist has written *urbem* instead of *orbem*, and the Clementine follows this tradition. So I have rendered, 'who turn the state upside down'; that is how the thing stands in every Vulgate in the world nowadays, and it is no part of the translator's business to alter, on however good grounds, his original.¹

Another important limitation imposed upon the translator is his use of the religious terminology. It is obvious that the community has during the centuries of its existence developed a definite method of exegesis. Some sentences have to be interpreted in a certain way to bring to light their several allusions which may be of great doctrinal significance. If the translator fails to make a rendering from which the same religious thoughts or tenets can be derived, if his translation may by implication mislead the reader, the Church will have to protect its lay community from heresy. The Church must forbid translations which may come into the hands of 'laymen and uneducated men and women' who would be unable to 'extract the true meaning'. Obviously the more rigid the exegesis of the Bible becomes, the greater the difficulty to make a new exegesis or a new translation. At the time of the growth of heretical sects the system underlying doctrinal thought may be so strong and the fear of heresy so alive that new translations may be forbidden. Two of the reasons which in the fifteenth century prompted the edicts against new versions into the vernacular

¹ *On Englishing the Bible*, pp. 1–2.

languages were the fear that the translator makes mistakes and that he cannot reproduce the full traditional meaning. This was at least partly due to the belief that the vernaculars were incapable of expressing 'what...writers have most accurately and sagaciously written on the deepest speculations of Christian religion and natural science', as was asserted in an edict of 1485/6.[1] The religious terminology in all its subtlety resulting from many generations of great thinkers who had played an important role in fashioning the tradition of the community was part and parcel of the exegesis. Although the Church may not be committed to any special philosophical system, the philosophical doctrines may contain some religious concepts which 'cannot, without impiety, be abandoned' without destroying theology itself. In Pius XII's Encyclical Letter of 1950 the importance of terminology has been clearly recognized:

> Treat with disrespect the terms and concepts which have been used by scholastic theologians, and the result, inevitably, is to take all the force out of what is called 'speculative' theology. It rests entirely on theological reasoning....[2]

Religious terminology must indeed be carefully retained by the religious authorities if they wish to keep the tradition alive. The translator must pay attention to this since many a word of the Bible has received its special significance within the system of theological reasoning.

Two examples may be adduced to show how the change of one word in a Bible translation may change the tradition. The

---

[1] For texts of the Edict containing Bible prohibition of Archbishop Arundel in 1408/9 see D. Wilkins, *Concilia Magnae Britanniae et Hiberniae* (London, 1737), vol. III, p. 317, para. vii, translation in J. Foxe, *Acts and Monuments*, vol. III, p. 245. The Act is quoted by B. F. Westcott, *A General View of the History of the English Bible* (London, 1905), p. 17. The edict of Mainz of 1485 is printed in V. F. Gudenus, *Codex Diplomaticus Anecdotorum...*, vol. IV (Frankfurt-Leipzig, 1758), pp. 469–71. An abbreviated edition is in C. Mirbt, *Quellen zur Geschichte des Papsttums und des römischen Katholizismus* (3rd ed. Tübingen, 1911), no. 332, p. 184. It is discussed by F. Kropatscheck, *Das Schriftprinzip der lutherischen Kirche*, vol. I (Leipzig, 1904), pp. 115 ff. A general discussion of the question of prohibition is found in, for example, G. Rietschel, 'Bibellesen und Bibelverbot' in Hauck, *Realencyclopädie*, vol. II, pp. 700–13, esp. pp. 703–4; M. Deanesly, *The Lollard Bible and other Medieval Biblical Versions* (Cambridge, 1920), pp. 18–88; H. Vollmer, 'Neue Beiträge zur Geschichte der deutschen Bibel im Mittelalter', *Bibel und Deutsche Kultur*, 8 (Hamburg, 1938), pp. 11 ff.   [2] *Loc. cit.* p. 10.

gospel of John starts with the words: 'In principio erat verbum....' When Erasmus of Rotterdam, relying on commentaries of the early Fathers of the Church, changed the word 'verbum' to 'sermo' in his translation of the New Testament, he implied that the scholastic theological speculation on 'verbum' was wrong since it was based on a wrong translation of the Greek Bible. It was not amazing that those who believed in the medieval philosophical tradition of the Church denounced the new translation.

The other example may prove how the change of terminology may endanger the very structure of the Church. The words in question were used by Tyndale and attacked by Thomas More in his *Dialogue Concerning Tyndale*. Thomas More asserts that Tyndale 'after Luther's counsel' has 'corrupted and changed' the Bible 'from the good and wholesome doctrine of Christ to the devilish heresies of their own, that it was clean a contrary thing'. The danger of Tyndale's version is, in More's view, that it 'was craftily devised like' the truth and that therefore 'the folk unlearned' cannot discern truth from falsehood in it. As a proof More discusses the translations of three words when Tyndale 'changeth the knowen usual names' of *charity* to *love*, of *priest* to *senior*, and of *church* to *congregation*. These changes are explained as 'mischievous' since they are due to Luther's teaching: *love* is used for *charity* because of Luther's tenet 'that all our salvation standeth in faith alone', and that good works are of no avail; *senior* has replaced *priest* because 'Luther and his adherents hold this heresy that all holy order is nothing'; *congregation* takes the place of *church*, for 'Luther utterly denieth the very catholic church in earth'.

Thomas More comes to the conclusion that Tyndale's translation, which contains more than a thousand errors, could not be corrected. It should be prohibited, which at that time meant it should be burned.

It is...to me great marvel [More writes] that any good Christian man having any drop of wit in his head, would anything marvel or complain of the burning of that book, if he knew the matter.[1]

---

[1] Book 3, chs. 8–10, quoted from the modernized version in *The English Works of Sir Thomas More*, ed. W. E. Campbell, vol. II, pp. 206 ff.

It is noteworthy that Thomas More has clearly pointed out the connexion between the interpretation of the Bible and the translation. He was moreover aware of the fact that the change of a word which has religious significance may constitute a break with tradition and may therefore mislead 'the folk unlearned'. He asserts that the change in terminology was intended by Tyndale under the influence of Luther as an attack upon the Catholic Church. Yet More was not opposed in principle to a translation of the Holy Writ into English but he demanded that such a rendering must be approved by the Church.[1]

It is impossible to say what principles underlay the first translations of the Bible before the authority 'guarding, preserving, and interpreting' the sacred writings could interfere with the translator. It is, however, known that even before the Christian era special consideration was given to Old Testament translation because of the holiness of its wording. It was thought that no human being is able to reproduce God's word as revealed to man in its entirety unless a new revelation takes place in which God makes known His word to the translator in a new language. If a new revelation occurs, the translation is equal to the original in every respect. The process of rendering is not executed with the help of human interpretation but through God's direct intercession. The translator is nothing but an instrument of God. This view of Bible translation will be called the inspirational principle of translation. Once an inspired version exists, its text is final, it cannot be revised or amended as long as no depravation of its actual wording occurs.

It is not my intention to discuss the existence or non-existence of divine inspiration in translation. From the historical point of view an existing translation may have been considered inspired and therefore identical with the original. If this claim is believed, if it is thought that an old version of the Bible owes its existence to inspiration rather than to human effort, this claim is equally solid and important whether it be true or not. When, as will be shown later, the Septuagint was believed to be of divine origin

[1] *Ibid.* Book 3, ch. 16, p. 242.

and the theory of inspirational translation was created, later translators had to take it into account. The creation of a new inspired version in its strictest form is almost a contradiction in terms, for in spite of his inspiration a translator must translate. A revelation in which man is only God's instrument can scarcely be a rendering of a given text. Yet it is possible to modify the theory and deny that a verbal inspiration of every word has taken place although the translation may still be inspired. Such a modification has indeed occurred during the history of European Bible translation.

Together with the inspirational principle another principle has always been in existence. Those who followed it did not believe in the existence of inspirational translation. They recognized and revered the mystery of God's word that cannot be fathomed by man but they thought that a translation can only be executed by the human mind. However imperfect man's understanding of Scripture may be, he may through study be able to arrive at a greater comprehension and thus produce a better translation which, however, can never be a reproduction of God's word in its entirety. This principle in its most radical form cannot therefore recognize any 'authentic' version that replaces the original and can be used as a basis of exegesis. This principle will be called the philological principle. Its rigid demands too may be modified to a certain extent and the gulf between the two principles may be narrowed, yet their basic conceptions are contradictory and therefore no complete harmony has been found between them.

A clash between these two principles occurred in the fifth century. When in changed circumstances the question was raised again in the sixteenth century, the two camps took part of their weapons from the armoury provided for them by the earlier controversy. This study deals mainly with the clash that took place in the sixteenth century. For its understanding, however, it is necessary to give a survey of the fifth-century dispute.

CHAPTER II

DISCUSSIONS ON THE ORIGIN OF THE SEPTUAGINT

The Septuagint is the Greek version of the Hebrew Old Testament; it was made in Egypt. The Pentateuch was probably translated in the third century B.C., the other parts of the Old Testament in later times. It would be important to know why the translation was made and what principles were applied in its execution. Unfortunately its origin is shrouded in mystery and later reports do not help us to discover the facts. These reports may record old traditions; but if so, the facts have been obscured by layers of additions through which original fact and legend have become inextricably entwined. It is not intended to analyse these reports here in order to bring to light these different layers in their development and ramifications or to unearth historical facts. In this chapter the story as handed down to us will be treated as a document that has its own value as a piece of literature and as evidence of trends of thought which were current at about 100 B.C., when it was written down. This so-called Aristeas Letter[1] describes how the Hebrew Old Testament, more exactly the Pentateuch, was translated into Greek. It purports to have been written by Aristeas at the time of King Ptolemy II Philadelphus (285–246 B.C.).

---

[1] *Editions*: *Aristeae ad Philocratem Epistula cum ceteris de Origine Versionis LXX Interpretum Testimoniis*, ed. P. Wendland (Teubner, 1900); H. G. Meecham, *The Letter of Aristeas*, Publications of the University of Manchester, no. 241 (Manchester, 1935) (Bibliography, pp. xiii–xviii).

Translations into English: R. H. Charles, *The Apocrypha and Pseudoepigrapha of the Old Testament in English*, vol. II (Oxford, 1913), pp. 94–122; *The Letter of Aristeas*, translated by H. St J. Thackeray (London, 1917), whose translation I generally follow; H. G. Meecham, *The Oldest Version of the Bible: 'Aristeas' on its traditional Origin*, the thirty-second Hartley Lecture (London, 1932) (Bibliography, pp. xvii–xxiii). The latest discussion on the LXX and the Aristeas Letter is by P. E. Kahle, *The Cairo Geniza*, The Schweich Lectures of the British Academy, 1941 (London, 1947), pp. 132 ff. One of the most valuable papers on the Aristeas Letter is by P. Wendland, 'Zur ältesten Geschichte der Bibel in der Kirche', *Zeitschrift für die neutestamentliche Wissenschaft*, vol. I (1900), pp. 267–90.

DISCUSSIONS ON THE ORIGIN OF THE SEPTUAGINT

The Letter begins with an account of an audience given by the king to his librarian Demetrius Phalereus, at which Aristeas was present. The king intended to collect all the books in the world in his library and Demetrius, telling him of his hope of having 500,000 volumes in his library shortly, gave his opinion that the Jewish Books of the Law should also have a place in it. During this audience Aristeas pleaded for the release of 100,000 Jewish slaves in Egypt, and this request was granted before any further steps were taken leading to the execution of the translation. A memorandum sent by the librarian to the king pointed out that the Books of the Law were written in Hebrew and that they ought to be translated anew; they were of a philosophical nature, but had not been mentioned by authors, poets, and historians because they contained a sacred and holy idea (§ 31). The king thereupon sent a letter to Eleazar, the High Priest in Jerusalem, and asked him to select seventy-two elders, six from each of the twelve tribes, 'men of noble life' (§39) who knew the Law and who were able to translate. This letter, together with gifts for the Temple, was taken to Jerusalem by the chief of the bodyguard and by Aristeas himself.

After a description of these gifts and an account of Jerusalem and Palestine, Aristeas relates how the high priest took leave of the seventy-two translators who, he feared, would be kept in Egypt since Ptolemy II was wont to summon all the wisest men to his court. After an eulogy on the Pentateuch by Eleazar, the scene shifts to Alexandria, where the seventy-two translators were received by the king with great honour. The king first paid reverence to the rolls of the Pentateuch which they had brought to Egypt, and then banqueted with them for seven days, asking every one of them questions about politics, military affairs, kingship, philosophy and so on. They answered all these questions to the greatest satisfaction of the king, who admired their wisdom. Three days later they started upon the translation of the Pentateuch, 'arriving at an agreement on each point comparing each other's work' (§302). The place where they gathered was 'delightful because of its quietness and brightness' (§307); every day before starting with their work the translators washed their hands in the sea and prayed to God. This was an action indicating that they had done no evil. In this way everything was directed towards righteousness and truth (§§305-306). They accomplished their task in seventy-two days 'as though this coincidence had been intended' (§307). The Jews were then called together, the translation was read to them, and

the priests and the elders of the translators and of the Jewish community[1] and the leaders of the people stood up and said, 'for-as-much as the translation has been well and piously executed and with perfect accuracy, it is right that it should remain in its present form and that no "revision" should take place' (§310).

A curse was pronounced upon anybody who should alter the text of this version (§311). The whole translation was read to the king, who was amazed that none of the historians and poets had mentioned these Jewish books which were so enlightened. At this point the beginning of the Letter is taken up where the librarian had drawn attention in his memorandum to the view that these books contain a sacred and holy idea. At the end of the Letter the librarian gave an answer to the king which was more explicit: these books are not mentioned, he said, 'because of the holiness of the Law and because of its origin by God' (§313). He then explained this somewhat obscure answer by saying that it was a dangerous thing to quote the Law and to reveal it to the people. When attempting to do so the historian Theopompos suffered a derangement of the mind which lasted for thirty days, and the poet Theodectes was afflicted with a cataract of the eyes. Both the historian and the poet were healed when they deleted the references to the Pentateuch in their writings (§§314-316). When King Ptolemy had learned this, he ordered that the Books of the Law should be kept with great reverence. The translators were given gifts and allowed to return home.

---

[1] For this interpretation of the text see R. H. Charles, *The Apocrypha*, vol. II, p. 121 (n. to 310).

The proceedings, so far as the Bible translation is concerned, are very similar to those of the twentieth century. Expressed in modern words they are: the highest dignitary of the religious community appoints a commission for the translation of the Bible. The members of this commission are of very noble character, they are scholars in theology, and they know the languages necessary for the translation (cf. Aristeas Letter, §121). They meet to compare their individual translations and to hammer out an agreement on the differences. This version, 'well and piously executed and with perfect accuracy', is acclaimed to be final. Nobody is allowed to reprint it in a corrected form, and that means that this version is, as it were, authorized. The writer of the Aristeas Letter comments on the imprecation uttered against anybody who might change the translation: 'And herein they did well, to the intent that the work might for ever be preserved imperishable and unchanged' (§311). In a modern report on a Bible translation this imprecation would naturally not be found and there would be no reference to the danger involved in a rash attempt to render Holy Writ. In the Letter precedents are mentioned warning people not even to quote the Law in any profane writing. There is another difference between modern procedure and that described in the Letter, namely the method of selecting the translators. In modern times the translators are known to be learned men capable of executing their task, in the Letter they must prove their worth in an examination to which they are subjected by the king during the banquets.

Whatever else may have been the intention of the author of the Aristeas Letter, his attitude towards Bible translation is clear. As it is assumed that it was written at about 100 B.C., we can glean very important information about the position of the Bible among the Jews in Egypt at that time: the translation of the Bible as a holy book has to be very carefully executed. For the making of an exact translation a commission is necessary which can discuss every detail. An agreement of the members of this commission which is bent on this holy task, is sufficient proof that the translation is accurate. This is the philological principle of translation. The translators themselves must be learned men and they must approach their work with a pious mind and be free from sin and therefore they purify themselves through washing their hands and through prayer. This, it seems, is not done to seek God's intervention in their task. In the Letter no miracle happens to enlighten the translators. The work of the human mind is sufficient to produce a translation of God's word. There is, however, one sentence which seems to point to some miraculous event. The number of the translators is seventy-two; they completed their work in seventy-two days 'as though this coincidence had been intended' (§307). The author, it seems, is playing with the idea that some miracle might have taken place or referring to another tradition about the origin of the translation.

This other tradition, which scholars consider to be of later origin than the Aristeas Letter, has been preserved by Philo of Alexandria and by the Fathers of the Church. There are differences within this tradition; for example, whether the translators worked separately in different cells without the possibility of communicating with one another or whether they worked together. But such disagreements are of no significance for this study. Only Philo's account will be discussed in detail, not merely because it is the oldest in this group but also because it sets out the principles of inspirational translation more clearly than any other source known to me.

From the very beginning, Philo with great literary skill creates an atmosphere removed from ordinary human life. Every action of the translators is filled with a significance which is important for the success of the rendering. Indeed, the translation is the centre which gives life to those who undertake this difficult task. Thus almost every word gains a new overtone within the narrative. It is impossible to reproduce these qualities of Philo's narration without quoting every word. However, there are more obstacles for the interpreter. By Philo's time neoplatonic philosophy and moral teaching had brought about semantic changes in many Greek words whose full significance it is not always easy to see. Philo's language, however, is even more complex because the connotation of his words is often coloured by both Greek philosophy and Jewish religious conceptions. This blend of different strata of civilization fully corresponds with Philo's attempt to create a union between Greek philosophy and Jewish religion.

This is Philo's account of the origin of the Septuagint:[1] The Jewish Law was, on account of its sanctity, marvelled at by Jews and other peoples. For a considerable time, 'while it had not yet revealed its beauty to the rest of mankind', it was in existence in Hebrew only. 'For a short time', Philo continues, 'envy causes the beautiful to be overshadowed but it shines forth under the benign operation of nature when its time comes' (II, v, 26–7). When the time had come for the Pentateuch to be known among the peoples, Philadelphus, the best of all kings of his own time as of the past, desired to have the Law translated into Greek. He sent messengers requesting the high priest of Jerusalem, who was also king of Judaea, to send translators to him; his wish was fulfilled and many translators arrived who were most esteemed and who had enjoyed an education in Hebrew and Greek. Philadelphus asked them difficult questions to test their wisdom (II, v, 29–33). This story, narrated at great length in the Aristeas Letter, is compressed into a few lines by Philo.

The translators then searched for a place where the rendering could be done. 'Reflecting how great an undertaking it was to render the Law which had been divinely revealed, and reflecting that they could not add or take away or transpose anything but had to preserve the original form of the Law and its character, they proceeded to look for a spot spiritually most pure, outside the city. For, within the walls, it was full of every kind of living creature, and consequently the prevalence of diseases and deaths and the impure conduct of the healthy inhabitants made them suspicious of it' (II, vi, 34). This thought, a perfect blend of

[1] *De Vita Mosis*, II, v–vii, 25–40. The text followed is that by L. Cohn-P. Wendland, vol. IV of Philo's *Opera* (Berlin, 1902). My translation is based on that of F. H. Colson, published in The Loeb Classical Library, *Philo*, vol. VI (1935). But I have changed this translation wherever I thought necessary without indicating my dissent.

Jewish and Neoplatonic ideas on ritual purity, causes the translators to move to the island of Pharos where even the roar of the sea is heard only dimly, in the remote distance. Here, they believed, they would find calm and quietness and their souls would be able to commune with the Law. They raised the sacred books heavenwards asking God to grant them that they might not fail in their purpose. God agreed to these prayers for the benefit of mankind (II, vi, 35-6). Now that God's blessing had been granted to this undertaking, the translators themselves were prepared, and their surroundings and their own disposition fitted to render God's Law. Here Philo's words may be quoted:

Sitting here in seclusion with none present save the elements of nature, earth, water, air, heaven, the genesis of which was to be the first theme of their sacred revelation, for the Law begins with the story of the world's creation, they [the translators] became as it were possessed, and, under inspiration, wrote, not each several scribe something different, but the same word for word, as though dictated to each by an invisible prompter. Yet who does not know that every language, and Greek especially, abounds in terms, and that the same thought can be put in many shapes by changing single words and whole phrases [or: 'by paraphrasing more or less freely'] and suiting the expression to the occasion? This was not the case, we are told, with this Law of ours, but the Greek words used corresponded literally with the Chaldean [i.e. Hebrew], exactly suited to the things they indicated. For, just as in geometry and logic, so it seems to me, the sense indicated does not admit of variety in the expression which remains unchanged in its original form, so these writers, as it clearly appears, arrived at a wording which corresponded with the matter, and alone, or better than any other, would bring out clearly what was meant. The clearest proof of this is that, if Chaldeans have learned Greek, or Greeks Chaldean [i.e. Hebrew], and read both versions, the Chaldean and the translation, they regard them with awe and reverence as sisters, or rather as one and the same, both in matter and words, and speak of their authors not as translators but as prophets and priests of these mysteries, whose sincerity and singleness of thought has enabled them to concur with the purest of spirits, the spirit of Moses.[1]

[1] *De Vita Mosis*, II, vii, 37-40.

Up to his own time, Philo concludes, a feast had been held every year on the island of Pharos to which Jews and many Gentiles came 'to do honour to the place in which the light of that version first shone out, and also to thank God for the good gift so old yet ever so young' (II, vii, 41).

In Philo's view the translation of the Hebrew Pentateuch is no ordinary rendering. It is not due to the endeavour of the human mind to transfer one language into another but to God's direct intervention. The translators cease to be translators, they are prophets and priests who are able to concur with Moses. Moses, however, is the 'purest of spirits', not a human being. The work of the translators was done under inspiration, they were mere instruments writing down their words 'as though dictated to each by an invisible prompter'. There was no necessity for them to discuss differences in their individual versions, for differences there were none. Their work was inspired and thus open to no error. The Pentateuch itself was the result of a revelation, as Philo points out (II, xxxv, 188), and only a new revelation can reproduce the exact wording of Holy Scripture as well as its form and character in a foreign idiom. The new version thus created cannot be compared with any other translation. It is in complete identity with the original, it is truly God's word.

The Greek Pentateuch is therefore the final translation which cannot be changed. Its authenticity is proven by its origin. Nobody who believes in the inspirational origin of the Greek text will attempt to make a new version in Greek. But it is this authenticity which set a new problem, when discrepancies between the texts were discovered: if both the Hebrew and the Greek texts are due to God's revelation, which is the original wording? Which version should be the basis for translations into other languages? This is an important question which was soon to become a burning issue in the Church.

We may round off this short account of the Aristeas Letter with some remarks on Philo's conception of prophecy, which is of importance for the later discussion.

DISCUSSIONS ON THE ORIGIN OF THE SEPTUAGINT

The highest function of the prophet is, in Philo's view, the proclamation of God's word. The prophet is an instrument only, he transmits to man what God has said to him. The prophet himself is not active, he is, as it were, only passive, an 'interpreter' (ἑρμηνεύς) of God's message. It is for this reason that the translators of the Pentateuch could be called prophets. But although praising this kind of prophecy in the highest terms (II, xxxv, 191), Philo maintains that the prophet should not be entirely passive, he should himself take an active part. This may be done in two different ways, and therefore there are in addition to passive prophecy also two kinds of active prophecy. In the one there is 'combination and partnership' (II, xxxv, 190) between God and the prophet. The prophet asks God for advice in a definite situation and thus, as it were, evokes God's help. In the other kind of active prophecy use is made of the gift of foreseeing, a gift which 'God has granted to the prophet' (II, xxxv, 190). The prophet, as a human being, sees, for example, a disaster approaching and then he is 'taken out of himself by divine possession' (II, xlvi, 250) and foretells what will happen. Although in these two cases the prophet is inspired by God, he is not a mere instrument for delivering His message. It is in this connexion that Philo says: 'interpretation and prophecy are of a different nature' (II, xxxv, 191). The meaning of the word 'interpretation' (ἑρμηνεία) includes 'translation'. Thus it could, in later times, be used to connote: 'Translation and prophecy are of a different nature'. In this form it will be found in the writings of St Jerome in his attack upon the theory of inspirational translation.

The philological and inspirational principles of Bible translation had been worked out before Christian renderings were made and before Christian theories on this subject could possibly have been forthcoming. These principles may have influenced the later versions of the Bible into Greek and Latin but many of them have been lost, and even where the names of the translators are known, none of them seems to have discussed these questions in theory.

DISCUSSIONS ON THE ORIGIN OF THE SEPTUAGINT

The references to the origin of the Septuagint in the works of the early Fathers of the Church prove that discussion continued on the questions set by the divergent views contained in the Aristeas Letter and in the works of Philo. It is important to take note of the fact that in the second century A.D. the Old Testament was translated into Latin from the Greek of the Septuagint and not from the Hebrew text. But who can say what was the reason for this? Was it considered that the Septuagint as an inspired version had replaced the original Hebrew text or was it due to the ignorance of the Hebrew language? In the third century A.D. Origen examined the relationship of the Septuagint to the Hebrew in his Hexapla. There he gave the Hebrew text and its transliteration in Greek characters, together with four Greek versions in adjoining columns. Moreover, he clearly marked additions to, and omissions of, the Hebrew wording in the different Greek versions. Although this laborious work was probably never copied in its entirety, it drew the attention of those interested to the fact that the Septuagint and the Hebrew original were not identical, as Philo had maintained. Yet the attestation of these differences, even if it were meant to refute the authority of the Septuagint, did not convince those who believed in the inspirational nature of the Greek version. Thus the divergencies of opinion remained, and at the end of the fourth century they led to an open controversy. The Latin Bible of the Latin-speaking Christian communities at that time was the *Vetus Latina* which, as mentioned above, was a translation from the Septuagint. The disagreement of its various manuscripts necessitated a reconstitution of the Latin text. St Jerome was called upon by Pope Damasus to undertake this task.

As far back as 380/1, before beginning the revision of the Bible, he had made some remarks on the Septuagint while speaking in general terms on translation. In the Preface to his translation of Eusebius' *Chronikon* he did so in a cautious and carefully worded way, not really coming to grips with the issue whether the Septuagint was inspired or not. But even in these short remarks

DISCUSSIONS ON THE ORIGIN OF THE SEPTUAGINT

he characteristically finds fault with the style of the translation, stressing the difficulties of a translator who wishes to preserve the elegance of the original without making additions.[1]

Style, indeed, was to be one point of departure for the attack upon the principle of inspirational translation. But before Jerome developed this side of his argument, he discussed the difficulties of Bible translation in the Preface to his revision of the New Testament of 384. In this he foresees that he will be censured by everybody as a falsifier and a sacrilegious person for changing words which everybody knew from earliest childhood. But it must be admitted, he says, that there are almost as many texts as manuscripts.[2] For the reconstitution of the text Jerome advocates the comparison of the Latin version with the original Greek text for the correction of all mistakes made by bad translators or careless copyists. The New Testament was, in his view, originally written in Greek with the exception of Matthew which he believes to have been composed in Hebrew and only later translated into Greek. While dwelling on the difficulties peculiar to the rendering of the New Testament, he mentions the Latin version of the Old Testament as being three degrees removed from the Hebrew original, since it was rendered from the Greek of the Septuagint, and not from the Hebrew text. He refuses to discuss which of the different existing Greek versions is correct. But he gives a rule for discerning the correct Greek text. 'The translation approved by the apostles should be regarded as correct', he writes.[3]

It is difficult to assess the significance of these words, in which Jerome expressly states that he does not wish to discuss the question of the Old Testament. The statement that the Latin is removed by three degrees from the Hebrew may be a statement of fact; it may, however, mean that the Septuagint is not an inspired translation which replaces the original. A confirmation of this latter view may be found in Jerome's doubt whether the

---
[1] P.L. vol. XXVII, cols. 35–6.
[2] 'tot enim sunt exemplaria quot codices.' P.L. vol. XXIX, col. 526.
[3] P.L. vol. XXIX, cols. 525–7.

Septuagint or any of the other Greek versions contains the correct rendering. He believes that the testimonies of the apostles will throw light on this question.[1]

It will be seen that all these arguments were again used by Jerome at a later stage of his life but that he gradually became more explicit and more outspoken in his views. The sayings of the apostles were to be of decisive weight in assessing the value of the Septuagint and the Hebrew text but they were not used, as they were at this early stage, to distinguish between the values of the translations. The question of the relation between the Hebrew original and the Septuagint came to the fore while the other Greek versions, important though they remained, proved to be of a secondary importance only. When he wrote the Preface to Eusebius' *Chronikon* in 380/1 and the Preface to the New Testament of 384 he may possibly not yet have seen the extent of the differences between the Hebrew Old Testament and the Septuagint. These prefaces were, in all probability, written at a time when he himself had not yet reached clarity in this matter. He was not yet able to judge, and it is a fascinating study to observe how every one of his points is slowly clarified and how the value of the Septuagint is more and more reduced until, at the end, he pronounced judgement against the Septuagint and against the theory of inspiration.

The first step to this end was the discovery of the serious differences existing between the Hebrew text and the Septuagint. This must have been the basis from which all the other results had to be derived. In the Preface to his revision of the Books of Chronicles written not before 395, Jerome discusses the divergencies of varying character and he enumerates the reasons for them. Some of the divergencies between the Hebrew and the Greek texts, he asserts, are due to copyists who, for instance, corrupted many of the Hebrew names. These are errors which

[1] *P.L.* vol. XXIX, col. 527: Neque vero ego de veteri disputo Testamento, quod a Septuaginta senioribus in Graecam linguam versum, tertio gradu ad nos usque pervenit. Non quaero quid Aquila, quid Symmachus sapiant, quare Theodotion inter novos et veteres medius incedat. Sit illa vera interpretatio, quam Apostoli probaverunt.

DISCUSSIONS ON THE ORIGIN OF THE SEPTUAGINT

should not be ascribed to the seventy translators 'who, filled with the Holy Ghost, rendered the truth'. The other discrepancies, however, go back to the translators themselves. These are additions found in the text of the Greek Septuagint only and not in the original Hebrew. They were made, Jerome maintains, 'partly for stylistic reasons, partly on the authority of the Holy Ghost'.[1] This is, at face value, a surprising and even contradictory statement. If the translators were filled with the Holy Ghost, as was asserted some lines earlier, they worked 'on the authority of the Holy Ghost'. But how does the conception of style come into this context? Jerome's words can be understood if it is seen that he had two different kinds of additions in mind. The one has no basis in the original Hebrew text. A word or a whole sentence found in the Septuagint corresponds to nothing of equivalent meaning in the Hebrew. I submit that in Jerome's view additions of this kind are made on the authority of the Holy Ghost and are thus inspired. But matters are entirely different when additions are made for stylistic reasons. In these cases, such differences between Hebrew and Greek as the lack of equivalent words and the presence of idiomatic and syntactical peculiarities forced the translators to add words in order to render the same sense. Additions of this type found in the Septuagint were made for stylistic reasons, they are not the result of inspiration. This method of translation, as Jerome had pointed out in the Preface to Eusebius' *Chronikon*,[2] exceeded the task of the translator.

This theory changes the central conception of inspirational translation. The translator, although inspired, is not like an instrument which is merely used by God to write down the single words and sentences. He does not, as Philo expounded, write as

[1] *P.L.* vol. XXIX, cols. 402, 404. Cf. L. Schade, 'Die Inspirationslehre des heiligen Hieronymus', *Biblische Studien*, vol. XV (1910), pp. 141 ff. Schade sees Jerome's development and he mentions the same examples from Jerome's works as I do. But he neither gives a close interpretation of these sentences nor does he indicate that every one of Jerome's statements on inspiration limits the field where inspiration could have taken place until, at the end, inspiration is entirely discarded.

[2] St Augustine held a similar view, see *P.L.* vol. XLII, col. 1068. For details see D. S. Blondheim, *Les Parlers judéo-romans et la Vetus Latina* (Paris, 1925), pp. cii ff.

if the words 'were dictated by an invisible prompter'. His inspiration, his being filled with the Holy Ghost, ensures that he renders the truth, but he has the liberty to choose the stylistic formulation of his translation. It is the translator, not God, who makes 'additions' to the original for stylistic reasons. He is inspired, but his inspiration is not verbal inspiration. His rendering contains the truth, yet it may differ from the original because of the idiomatic peculiarities of the individual languages. In Philo's account of an inspired version the human element is completely denied, in Jerome's conception it is allowed to find a foothold again; the human brain is not overpowered, but works even during the inspiration.

From this the conclusions for the later translator are obvious: the *contents* of both the original and the inspired translation must be taken into consideration at every step, but so far as the wording is concerned, it is only the original which is binding. The inspired version thus complements but no longer replaces the original.

This interpretation of St Jerome's passage not only dispels any contradiction which might be contained in his Preface to the *Paralipomena*, it also fully agrees with his views on prophecy. The prophet has, according to St Jerome, the faculty to speak and to be silent, he does not speak against his will ('invitus'). He is inspired but retains his intellectual faculties and knows what he says.[1] In the same way the inspired translator knows what he writes and formulates it according to his own judgement.

The above interpretation is also consistent with Jerome's work at this period of his life. He uses the Hebrew text together with that of the Septuagint and informs the reader of the differences between these texts through diacritical signs without rejecting any of these divergencies. Both texts are of value and both must therefore be used together.

Further examination and comparison between the Hebrew and Greek texts may have forced him to modify this position. While working on the revision of the Bible he came to the conclusion

---

[1] *P.L.* vol. xxv, col. 1274B; cf. L. Schade, *loc. cit.* pp. 21ff.

that at some places the original Hebrew offers a better wording than the Greek translation. In the Preface to his *Hebraicae Quaestiones in Genesim* of between 389 and 392 he criticizes the Septuagint, asserting that the Seventy translators concealed from King Ptolemy II the meaning of all those passages in which the coming of Christ was promised. A further criterion that could be used to discover the reliability of the Septuagint is contained in the New Testament. According to St Jerome, the apostles' quotations of the Old Testament are testimonies for the soundness of the Hebrew text against the wording of the Septuagint. The conclusion is that an agreement between the Hebrew Bible and the New Testament is a full proof of the authenticity of the Hebrew text and a condemnation of the Greek version whenever the Septuagint disagrees. To this argument Jerome adds that the Seventy translators rendered only the Pentateuch, and not the whole of the Old Testament.[1]

These observations are of the utmost importance for his assessment of the value of the Septuagint for the rendering of the Bible. Jerome does not expressly say that the translators were not inspired. As only the Five Books of Moses were rendered by them, the translation of the other parts cannot claim the same authority. But even in the Pentateuch the intention of the translators played a great part in making the rendering, since they left out passages which should, in their time, not be divulged. The critic is able to discover the purpose of these translators. Their design, praiseworthy though it had been, was destined for their own time and is no longer valid. All traces of their activities had therefore to be removed and to be replaced by that text of the Bible which had been revealed to man in Hebrew.

This is the end of a long development in Jerome's thought. It is difficult to say how much his earlier statements were influenced by considerations of expediency and caution. It is well known that he was heavily attacked by all those who revered the Septuagint and who valued the *Vetus Latina* which had been

[1] *P.L.* vol. XXIII, cols. 936–7.

derived from it.¹ But Jerome was not discouraged in pursuing his way after he had arrived at a full comprehension of the facts. He followed his own investigation and did not hesitate to publish the result of his study. It is the philologist's method to compare the different texts and to rely on the ability of human understanding to find out the truth. In this research there can be no halt. When after a long period of uncertainty he at last found what he believed to be the truth, he drew the logical conclusion, even when this meant a fight against a long tradition and against strong opposition to all new ideas and thoughts.

Jerome had discarded the view that the text of the Septuagint was a faithful translation of the Hebrew original. He no longer believed that the Greek version was inspired, and for this reason identical with the Hebrew text despite differences of detail. This new recognition led necessarily to two conclusions. One was practical: a new Latin translation from Hebrew had to be made. The other was theoretical, derived from the belief in philology: inspirational translation does not exist. 'One thing I know,' he wrote, 'I could translate only what I had understood before.'² Translation is based on the comprehension of the original and the command of languages. It is not prophecy. Thus Jerome in his condemnation of the inspirational principle of translation turns Philo's sentence that 'Translation and prophecy are of different nature' against Philo's view about the inspirational origin of the Septuagint.³ And mockingly he adds that the assumption of the divine revelation of the Septuagint makes it imperative to argue that Cicero's translations, which were rhetorical, were inspired by the rhetorical spirit. In addition, he repeats that the apostles' sayings do not agree with the Septuagint. The inspiration of the apostles is, of course, beyond a shadow of doubt. If the Septuagint

---

[1] Testimonies are collected by Ferd. Cavallera, *St Jérôme sa vie et son œuvre* (*Spicilegium Sacrum Lovaniense*, Études et Documents, fasc. 2) (Louvain–Paris, 1922), vol. II, pp. 103–15. [2] *Preface in Job* (P.L. vol. XXVIII, col. 1081A).

[3] It is not known whether Jerome was consciously quoting Philo, but I am sure that Philo was the ultimate source. About Jerome's knowledge of Philo see P. Courcelle, *Les lettres grecques en occident de Macrobe à Cassiodore. Bibliothèque des Écoles Françaises d'Athènes et de Rome*, fasc. 159 (2nd ed. Paris, 1948), pp. 70–1.

too were inspired, the Holy Ghost would have made conflicting statements at different times. This is impossible and thus the Septuagint is not inspired.

These arguments are propounded in the preface to Jerome's new translation of the Pentateuch. He gathers together all the proofs he can find to defend his position and to show the necessity for a new translation. But he also adduces evidence against those who believe in the supernatural origin of the Septuagint. This belief, he points out, cannot be supported by any historical documentary evidence. The cells in which the Seventy translators are supposed to have done their work separated from one another were built by a liar. It is mentioned neither in the Aristeas Letter nor in Josephus. Thus the report about the inspiration of the translators is a legend supported by neither historical evidence nor theological reasoning. Some of Jerome's words may be quoted here:

I do not know [he says] who was the first who through his lies built seventy cells in Alexandria in which they [the translators] were separated and yet all wrote the same words; whereas Aristeas...and long after him Josephus have said nothing of the sort, but write that they were assembled in a single hall and conferred together, not that they prophesied. For it is one thing to be a prophet and another to be a translator; in one case the Spirit foretells future events, in the other sentences are understood and translated by erudition and command of language.[1]

St Jerome's principle of translation was at this time firmly established in his mind. The Septuagint may be used as any ordinary translation but it cannot replace the original Hebrew. The Latin text of the Old Testament and of the New Testament should be treated in the same way, if the readings in the different manuscripts differ. The correct text will be established through comparison of the Latin translation with the original Hebrew of the Old Testament and the Greek of the New Testament. 'One must go back to the Hebrews', he writes, 'from whose midst

---

[1] *Praefatio in Pentateuchum* (P.L. vol. XXVIII, cols. 150–1); cf. *ibid.* col. 1081.

speaks even the Lord.'[1] Thus the Hebrew text had come back into its own.

The method of Jerome's translation is a matter of considerable dispute. Conflicting statements have been made by Jerome, and there is no possibility of harmonizing them or of proving that he never changed his mind. All his sayings from the very beginning show, beyond all doubt, his interest in style. It will be remembered that in his view lack of elegance in the Septuagint induced later translators to make new renderings of the Bible into Greek. It may be recorded that in his letter to Sunnia and Fretela he wrote that the task of a good translator consists in rendering idiomatic expressions of one language into the modes of expression peculiar to the other.[2] It may be mentioned that in some prefaces of his Bible translation he emphasizes that he does not render word for word, but sense for sense, while in others he maintains that he translates partly the words, partly the sense.[3] Indeed it seems that he made no difference between a translation of a profane book and one of the Bible. From the outset he had proclaimed that the singular characteristics of languages often force the translator to render the sense instead of the word. Even Homer, he points out, would sound incoherent if rendered word for word without regard to metre and contents.[4] Yet at the end of 395 he wrote in a letter that he had always rendered the sense 'with the exception of Holy Scripture where even the order of words is a mystery'.[5] In these words Jerome clearly advocates two different methods of translation. A literary work, he asserts, must be translated according to sense, as the lack of equivalents in the two languages and the preservation of the order of words make the word-for-

---

[1] *Praefatio in Paralipomena* (P.L. vol. XXVIII, col. 1326A): 'Ad Hebraeos igitur revertendum est unde et dominus loquitur.' *Ep.* to Sunnia and Fretela of uncertain date between 393 and 401, *Ep.* CVI, 2, 3 (P.L. vol. XXII, col. 838).

[2] *Ep.* CVI, 3, 3 (P.L. vol. XXII, col. 839).

[3] P.L. vol. XXVIII, col. 1433; P.L. vol. XXIX, col. 39.

[4] *Ep.* LVII, 5, 6–8 (P.L. vol. XXII, cols. 570–5). No details are necessary since this question has often been discussed. The latest study on this subject is found in P. Courcelle, *op. cit.* pp. 42 ff., where also see for literature.

[5] *Ep.* LVII, 5 (P.L. vol. XXII, col. 571).

word translation sound ridiculous and incoherent. Such a method destroys all literary qualities.[1] It is therefore necessary to render the sense in its entirety though not the words. In the Bible, however, every word is sacred. 'The revelation of John contains as many mysteries as words', he wrote in a letter of 395.[2] The meaning of the sacred text cannot be exhausted; it is like the ocean, inexhaustible and mysterious.[3] This mystery must be preserved in the translation. As the order of words transcends human understanding, a change in the order of words would not only destroy this mystery but it would also endanger the unfathomable profundity of the sacred text.[4]

It is impossible to know why Jerome advocated the word-for-word method of Bible translation while he himself did not follow it. For this statement is inconsistent with his theory and with his practice. The student always likes to see the person whose activities he is investigating consistent in all his utterances and deeds. It is only too easy to forget that even a Father of the Church was a human being, perhaps irascible and irritated by frequent attacks against him. In such moments he could easily write something contradicting earlier statements without being conscious of any change in his attitude. He may think of a certain example which at a certain moment he makes the pattern for a theory. Inconsistencies and even contradictions sometimes reveal the humanity of an author which it may be difficult to find in his other writings. Documentary evidence cannot, of course, be offered for such explanations, and this is the reason, I believe, why these inconsistencies are confusing. It may be suggested that Jerome's demand for a word-for-word translation is the result of a controversy and that it was written in the heat of it. It is known that the letter in which these words occur was provoked by Rufinus's

---

[1] *Ep.* LVII, 5, 6–8; *ibid.* 6, 2 (*P.L.* vol. XXII, col. 572). Cf. the Preface to his translation of Eusebius, *Chronikon* (*P.L.* vol. XXVII, cols. 34–5).

[2] *Ep.* LIII, 8, 6 (*P.L.* vol. XXII, cols. 548–9); cf. *Ep.* LVII, 11, 4 (*P.L.* vol. XXII, col. 578). For the date see F. Cavallera, *op. cit.* vol. II, pp. 89–91.

[3] *Comment. in Abacuc*, II, 3 (*P.L.* vol. XXV, cols. 1317–18).

[4] *Ep.* LVII, 5, 2 (*P.L.* vol. XXII, col. 571). Cf. W. Schwarz, 'The Meaning of *Fidus Interpres* in Medieval Translation', *Journal of Theological Studies*, vol. XLV (1944), p. 75.

attacks against him.¹ The prefaces of his Bible revision and his Bible translation prove that from the beginning he was subjected to heavy attacks from all sides. And he felt that all this censure was unjustified. He did not hold himself to be a falsifier of the Bible when correcting the old version or when making a new translation but rather a corrector whose intention it was to produce a faithful text of the Bible. This work, he emphasized over and over again, did not imply the condemnation of earlier translators.² But he wished to end a state of affairs where, owing to different readings, there were as many different texts of the Bible as there were manuscripts. He was very conscious of the tendency to retain an old text even though it was faulty and to condemn those as falsifiers who attempted to correct it.³ Consideration for this attitude may well have been the reason for retaining the text of the earlier Latin version, although he disagreed with the rendering and would have preferred a change.⁴ Attacks against him taught him (if he had not been aware of it before) the restrictions to which a translator of Holy Writ has to submit. Sometimes, at least, he gave a literal rendering although, as has lately been shown, the most literal renderings in his version were taken from earlier versions.⁵ Thus, I submit, one can understand that writing on his Bible translation at a time when he was censured by everybody, he made this statement on literal translation and was convinced that he was correct. For he had, after all, renounced a great part of the ornamentation of style and paraphrase which he employed in his renderings of profane works.⁶

¹ See, for example, G. Bardy, *Recherches sur l'histoire du texte et des versions latines du 'De Principiis' d'Origène* (*Mémoires et Travaux* fasc. 25) (Paris, 1923), pp. 161-3.
² E.g. *P.L.* vol. XXVIII, cols. 463, 1079, 1082A.
³ *P.L.* vol. XXIX, cols. 525-6; *P.L.* vol. XXVIII, cols. 147-8. Cf. A. Hauck, *Realencyclopaedie*, s.v. 'Bibeluebersetzung', pp. 36ff.; G. Bardy, *loc. cit.* p. 167.
⁴ E.g. *P.L.* vol. XXIX, col. 528 (*Ep.* CVI, 12); *P.L.* vol. XXII, col. 843.
⁵ G. Cuendet, 'Cicéron et saint Jérôme traducteurs', *Revue des Études Latines*, vol. XI (1933), p. 383.
⁶ G. Cuendet, *op. cit.* pp. 384-6, 387 ff. F. Blatt, 'Remarques sur l'histoire des traductions latines', *Classica et Medievalia*, vol. I (1938), pp. 219ff. P. Courcelle, *op. cit.*, disagrees with Blatt (p. 45, n. 5), but (pp. 45-6) he seems to come to Blatt's conclusion. Nobody has seen that there may be a connexion between this method and Jerome's words in *Ep.* LVII. Cf. G. Bardy, *op. cit.* pp. 158ff., 163.

Jerome's importance for the history of Bible translation is twofold. His advocacy of a word-for-word method of translation for the Bible was, I believe, accepted almost without exception during the Middle Ages. Moreover, the philological principle (as opposed to the inspirational principle) was recognized as being the basis of every translation. His Bible translation was destined to become the authentic version of the Church. His prefaces to the individual books of the Bible were often copied together with his version. Thus it was generally known that 'it is one thing to be a prophet and another to be a translator'.

The inspirational principle found a defender in St Augustine, who took up the challenge contained in Jerome's work.[1] As early as 394 he was opposed to a new version of Holy Scripture unless St Jerome added the diacritical signs for the indication of all the differences found in the Septuagint and the Hebrew texts. He emphasized the authority of the Septuagint whatever its origin, pointing out that he was at a loss to understand how anything could be found in the Hebrew text that had escaped the attention of all the earlier and very learned translators.[2]

In 403, however, Augustine was more outspoken. In the meantime, Jerome had rendered part of the Old Testament and this new version had been read in some churches. Augustine had grave misgivings about this practice. What would the religious communities say, he asks, if they found out the discrepancies between the Bible used before, and Jerome's new version? In North Africa, where there were many Greeks, it would be noticed that Jerome's text disagreed with the Septuagint. But the Hebrew text, from which the new rendering was derived, would not and could not be consulted by those who did not believe in the correctness of the new text. Even if they compared the Hebrew original with Jerome's version, 'who would condemn so many Latin and Greek

[1] A good bibliography is found in P. Courcelle, *op. cit.* pp. 401–20. Courcelle (pp. 144–53) often mentions Augustine's view on translation but without considering *De Civitate Dei*, XVIII, 42, where the inspirational method is discussed.

[2] Augustine, *Ep.* XXVIII, 2, 2 (*P.L.* vol. XXXIII, col. 112).

DISCUSSIONS ON THE ORIGIN OF THE SEPTUAGINT

authorities?' Augustine then continued with a description of disturbances which had taken place at Tripoli (Oea) when the text of Jonah was read in Jerome's new wording. This new version, Augustine maintained, was different from that text which 'was rooted in the affection and memories of all the people [sensibus memoriaque inveteratum] and which had been read during so many succeeding generations'. The Greeks, Augustine records, took an especial part in these disturbances accusing the new wording as being false. The bishop was forced to denounce the new translation fearing that he would be left without a congregation.

Obviously this incident shows the insistence of the people upon the traditional text of their Holy Book. But as this description of the disturbance is coupled with Augustine's opposition to Jerome's translation of the Old Testament it reveals Augustine's fear lest this new text might create a cleavage in Christendom. He clearly saw the dangers arising from Jerome's undertaking. He knew of the discrepancies between the Hebrew text and the Septuagint. But these divergencies had been of little account as long as the Church used one text only, namely the Septuagint and the Old Latin version derived from it. Jerome's new translation of the Old Testament changed the position of the Septuagint through the creation of another text which was in disagreement with the version used until then. The question might easily arise which of the two texts was correct. Would not the acceptance of Jerome's text involve the making of a new translation from Hebrew into Greek? Considerations such as these might underlie Augustine's desire to preserve the Septuagint against all attacks and to ask the question: should one condemn so many Greek and Latin authorities?

Thoughts like these may have been the reason for Augustine's opposition to Jerome's work on the Old Testament, for he is pleased with Jerome's emendations of the New Testament, since, as he points out in the same letter, these are no 'stumbling block'. Anyone can verify the text from the Greek original or, if he has no knowledge of Greek, can easily be instructed. Concluding this letter, Augustine admits that the Latin manuscripts of the Bible abound in different readings and proclaims that Jerome would be of the greatest use if he reconstituted from the Greek a faithful Latin text of 'that part of Scripture which the seventy have translated'.[1]

The polarity between the views of Augustine and Jerome can scarcely be seen in a clearer light. Augustine was not convinced by Jerome's attack on the Septuagint. He could not understand how Jerome could expect to elucidate the meaning of the Hebrew Bible after the attempt of so many learned people before him. 'If the text is obscure, you too will probably be mistaken; if it is clear, it is incredible that the earlier translators were mistaken', he wrote to Jerome in 394. As to the Septuagint, its authority is too great for one man to oppose it.[2]

At first sight two quite separate motives seem to be operating in Augustine's mind, one based on theology, one on Church policy. Which of these should be regarded as primary? Was it belief in the Septuagint and mistrust of man's critical capacity or was it the authority of the Church whose tradition had to be upheld? How could Jerome be more learned than those other than the Seventy who had translated the Bible into Greek? How could he avoid mistakes? Or was it fear lest the unity of Christendom would be destroyed if the old Latin Bible was replaced by a rendering from Hebrew? No positive answer can be given to these questions. Indeed, a closer study of Augustine's thought makes it seem unlikely that he would have considered these two approaches to be incompatible. Any dichotomy can be resolved by the assumption of one firm principle underlying his attitude to this problem. The Septuagint is a faithful translation of God's word, for it is inspired. Only when his belief in the inspirational nature of this translation is taken into account, is Augustine's view of translation coherent and logical in all its details. As far as is known to me there is no recent study on Augustine's thought on

---

[1] *Ep.* LXXI (P.L. vol. XXXIII, cols. 241–3).
[2] *Ep.* XXVIII, 2, 2 (P.L. vol. XXXIII, col. 112).

DISCUSSIONS ON THE ORIGIN OF THE SEPTUAGINT

translation that mentions the inspirational principle of translation. Indeed it is typical of the rationalism of the nineteenth and twentieth centuries that it has been forgotten.

Augustine expressed his view on the Septuagint twice, in his *De Doctrina Christiana* of 416/19 and in the *De Civitate Dei* of 410/28.[1] The argument in these two works is the same and has the same purpose: to prove the divine origin of the Septuagint. In both he attacks Jerome's version, but mentions his name only in the *De Civitate Dei*.

Augustine speaks of the miracle of the Septuagint. Although separated from each other, the translators found that all their versions completely agreed in every word and in the order of words. 'There was, as it were, one translator only, for [*ita*] the translation of all of them was one and the same; indeed, one spirit was in all of them', he reports in the *De Civitate Dei*. Because of this miracle, he concludes, the authority of the Septuagint is not human but divine. Augustine follows the tradition which had been so eloquently expressed by Philo. This tradition exalts the importance and excellence of the Septuagint high above all the other translations of the Old Testament. Following this view Augustine is able to dismiss the other versions with a few words and to point out that the Greek and Latin Christians have accepted the Septuagint, the Greeks in its original language, the Latins in a translation, the old Latin version. The decision of the Church to use a text derived from Greek, and not from Hebrew, is thus fully justified. It is necessary to stress this point, for otherwise it escapes the reader that the words following contain a severe censure of Jerome's version from Hebrew:

However, in our time there was a priest Jerome, a very learned man, knowing all three languages, who translated these Scriptures not from Greek, but from Hebrew into Latin.

The Jews, Augustine goes on, hold that Jerome's is a faithful rendering while the Septuagint is not. Yet the Church should

[1] I follow Courcelle's dating of *De Doctrina Christiana* (*loc. cit.* pp. 149–50). It was generally dated 397.

prefer a work made by so many excellent men to any version executed by one person only.

Having mentioned Jerome, Augustine had to defend his view against all those who did not believe in the inspiration of the Septuagint. It will be remembered that Jerome attacked this assumption on two accounts, first, the translators did not work separately, and secondly, they made omissions and additions according to their own lights. Assuming, answers Augustine, that the translators were not guided by the Holy Spirit and that they had, after a comparison of their versions, decided upon a text, this their agreement is still weightier than the translation of any one person:

> But [he continues] as the sign of divinity was truly manifest in them, it follows that every other translator of those Scriptures from Hebrew into any other language is right only, if he is in agreement with those Seventy translators. If he be not, one has to assume that it is they who are gifted with the true depth of prophecy. For the Spirit that was in the prophets when they spoke, this very Spirit was in the Seventy men when they translated.[1]

From the divergencies between the Hebrew original and the Greek text Augustine draws conclusions contradicting Jerome. It must have been the Holy Spirit who with God's authority changed the words though not the sense. Additions and omissions were made in the Septuagint to prove that 'no human servitude to the words was at work which the translator ought to have but rather a divine power which filled and ruled the translators' minds'. It follows that

> whatever is found in Hebrew and not in the Septuagint, this the Spirit of God wished to say through those prophets and not through these. Conversely, what is found in the Septuagint and not in Hebrew, this the same Spirit intended to say through these prophets rather than those, manifesting in this way that both were prophets.

[1] *De Civitate Dei*, XVIII, 42-3 (*P.L.* vol. XLI, cols. 602-4).

The Holy Spirit, Augustine continues, did indeed not say the same to every one, as witness the fact that he spoke differently to Isaiah and to Jeremiah and to the other prophets. Yet where the prophets and the translators of the Septuagint agree, there the 'one and the same Spirit' wanted both of them to say the same, 'in such a way that the prophets preceded them in time with their prophecies, and that those who made the prophetical translation followed them. For as there was one Spirit of peace among those who spoke true and concordant words, so the same one Spirit was apparent in those who without conferring among themselves nevertheless translated, as if with one mouth.'[1]

Augustine had thus answered Jerome's attacks upon the Septuagint and imbued it with all the qualities found only in an inspired translation. The prophet through his rendering makes a version which replaces the original. This translation must therefore be accepted as God's Word and Jerome's Latin rendering must be wrong, since it was made from the original Hebrew which had been superseded by the new inspired version. From this two conclusions must be drawn: if the Latin text of the Old Testament is corrupt, it must be emended from the Greek of the Septuagint.[2] The other conclusion is that the ordinary translator, unlike the prophetic translator, must be in 'servitude to the words', a phrase found in the *De Civitate Dei*.[3] Augustine must therefore be in favour of the word-for-word technique of translation. A translator who follows the sense often changes not only single words but also whole phrases, he maintains. In the attempt to render idiomatically he often departs from the meaning of the original. A translator, however, who renders word for word retains the meaning of the original. His style may contain barbarisms and solecisms which should be avoided if they lead to ambiguities.

---

[1] *De Civ. Dei*, XVIII, 43 (*P.L.* vol. XLI, col. 604); cf. *De Doctr. Christ.* II, 15 (*P.L.* vol. XXXIV, col. 46).

[2] *De Doctr. Christ.* II, 25 (*P.L.* vol. XXXIV, col. 46).

[3] *De Civ. Dei*, XVIII, 43 (*P.L.* vol. XLI, col. 604). I wonder if there is any connexion between this phrase and Origen, *Epist. ad Afric.* 2, directed against Aquila: δουλεύων τῇ Ἑβραϊκῇ λέξει (*P.G.* vol. XI, col. 52B).

But if the original can be rendered more faithfully by the use of a style which is not literary, this should be done. Ambiguities should be avoided at all costs.[1] Purity of style is of no importance for him who wishes to recognize the real contents and not only letters and words. Augustine's words are:

The weaker man is, the more he is offended by impurity of style, and the weaker he is, the more learned he wishes to appear, learned not in the knowledge of those matters which edify man, but in the knowledge of signs only, which most easily inflame the mind, for even the knowledge of those matters which edify man often lifts up man's neck unless it is curbed by the yoke of the Lord.

Augustine elaborates this theme until at the end he quotes: 'Because the foolishness of God is wiser than men; and the weakness of God is stronger than men.'[2]

Here then Augustine emerges as mistrusting human capacity for understanding. It is God who must lead man. It is God who has revealed the Bible twice, to the prophets and to those who, by inspiration, translated the earlier revelation into Greek. It is foolish for a human being to reject this conception of inspirational translation and to undertake a translation of the Bible relying on man's abilities. Even at its best the work created by man must be inferior to the version of the prophet. Only the inspirational translation is binding. It replaces the original, for it is truly God's word.

What was the outcome of this discussion between these two Fathers of the Church whose different points of view reflect not only the differences in the thoughts of these two great personalities but also the clash between inspirational and philological principles? It is obvious that no final answer is to be expected in which either of these theories is condemned. Rather is a solution based on expediency given from time to time. This bridges the gap for

---

[1] *De Doctr. Christ.* III, 2–3 (*P.L.* vol. XXXIV, cols. 65–9). For details see Courcelle, *op. cit.* pp. 148–9.

[2] *De Doctr. Christ.* II, 13, 19–20 (*P.L.* vol. XXXIV, cols. 44–5); 1 Cor. i. 25.

some time, but is insufficient to prevent another clash from occurring later. In the fifth century the question, it seems, soon lost its importance. In spite of St Augustine's opposition, St Jerome's translation was soon used in churches and thus became the authentic text of the Church.[1]

---

[1] The Vulgate was officially authorized only by a 'decretum' of 8 April 1546. The reason given then was that the Vulgate 'longo tot saeculorum usu in ipsa ecclesia probata est'.

CHAPTER III

THE TRADITIONAL VIEW

Jerome's translation was the authoritative Bible of the Western Church during the Middle Ages even though it was not officially recognized. It was used everywhere. Missionaries took it with them, and whenever a translation into the vernacular was made, it was this Latin text which served as the original. The ignorance of Hebrew and Greek in Western Europe helped to establish this supremacy. The fact that the Vulgate replaced all other texts of the Bible is of the greatest significance for the history of Bible studies and of Bible translation, since nearly all interpretations and commentaries were based on the Latin text without regard to the wording in the original languages.

The medieval exegesis was, to a very large extent, incorporated into the teaching of the Church. Only the Church had the right to decide whether any view in matters of faith was correct. This meant, in theory, that any interpretation not objected to by the Church was considered to be valid and binding on every member of the Christian community. The interdependence of the text and its interpretation had important results. For any change of the text (assuming that the wording were found to be corrupt) might easily nullify the accepted exposition and thus lead to the belief that the Church had erred in allowing a certain interpretation. The hostility against any attempt to change the text of the Vulgate, or even to restore it to its original form through emendations, became gradually stronger over the centuries. It gained considerable reinforcement from the great philosophical edifices which brought into agreement all the sayings of the Bible and all the divers interpretations of the Fathers of the Church.

The translator of the Bible has to respect these forces. He has to decide whether to reject or to follow the authoritative wording and its interpretation, he has to make up his mind as to the wisdom of making a new version in a vernacular which may involve defamation and Church ban if it is found wanting. His decision to reject or to follow a certain exegesis is thus dependent not only on his own personal view but also on the validity and strength of the authoritative version. The student of Bible translation must therefore try to view every version in the perspective of the religious tendencies of the time of its origin. It has been shown in the preceding chapter that St Jerome had to overcome the opposition of those who defended the tradition in its entirety in spite of obvious corruptions of the text. But he was also opposed by those who rejected philology as a method for establishing the meaning of a divinely inspired text. In the sixteenth century this controversy was to arise again. To understand its significance, one must ask the preliminary question: What was the validity of the Vulgate before the controversy began and before the new translations were made?

Such a question cannot be answered with one clear-cut statement. Different people naturally followed different currents of thought so that a complicated and often perplexing picture is presented to the observer. In the fifteenth and sixteenth centuries three approaches to this question can be detected: the acceptance of the whole tradition in its historical development; the acceptance of the oldest tradition only, later development being rejected; and lastly the complete denial of any authoritative tradition. I shall call the first of these currents of thought the 'traditional view', the second the 'philological view' and the last the 'inspirational view'.

It is the nature of the traditional view to be static, preserving the past and giving way to new thoughts and ideas only when this is unavoidable. It follows the old, approved method of Bible interpretation. Even if one or the other point is stressed more than it had been done before, the guiding principle is the conservation of the permanent tradition. Its strength lies in its long history, which goes back to the Fathers of the Church and thus directly to

THE TRADITIONAL VIEW

God's divine inspiration.[1] With this strength, however, is bound up an integral weakness: the system of thought erected during the Middle Ages became so closely integrated that there was no room for new thought, since this would inevitably conflict with the tradition in some way. New works therefore were mostly written to clarify and amplify points made before.

The method of Bible exegesis was evolved through long experience. Its basic principle was the fourfold sense of the text: the literal sense explains the historical contents, the allegorical sense elucidates matters of faith by revealing the allegory implied in the Biblical text, the moral sense indicates rules of human conduct, whilst the anagogical sense deals with the future to be hoped for. This method was still alive in the sixteenth century, as can be learned from the fact that it was, for example, mentioned in the preface to the famous polyglot Complutensian Bible, printed at Alcala from 1514 to 1517 and published in 1520. In its first volume the following old verse is quoted:

> Littera gesta docet: quid credas allegoria.
> Moralis quid agas: quo tendas anagogia.[2]

It goes without saying that not every sentence of the Bible was assumed to contain all these four senses.

The method used may be elucidated by a description of a typical print of the Bible as published at the end of the fifteenth century, the commentated Latin Bible published at Basle in 1498 and re-published in 1502.[3]

---

[1] Some material for a history of interpretation is found in B. Smalley, *The Study of the Bible in the Middle Ages* (Oxford, 1952), pp. 1ff. It is not my intention to go into details of this question or to discuss F. W. Farrar, *History of Interpretation*, the Bampton Lectures of 1885 (London, 1886). His general attitude based on the rationalism of his time differs in every respect from the point of view exhibited above which, I assume, is influenced by twentieth-century thought. The only recent book discussing the whole subject is P. C. Spicq, *Esquisse d'une histoire de l'exégèse latine au moyen âge*, Bibliothèque Thomiste, 26 (Paris, 1944).

[2] Fol. vi^r.

[3] The exact dates are 1 December 1498 and 15 May 1502. The Bibles were published by Johannes Petrus and Johannes Froben, Basel. The bibliographical details are discussed in the *Gesamtkatalog der Wiegendrucke* (Leipzig, 1925–), s.v. 'Biblia', no. 4284. Cf. R. Proctor, *An Index to the early Printed Books in the British Museum* (London, 1898),

Even the set-up of the pages reveals the traditional principle. In the middle of each page there is the text of the Bible in rather large letters. Between these lines the interlinear gloss is printed in small letters. These Bible verses are surrounded by the following commentaries:

(1) The ordinary gloss.
(2) The *Postillae* of Nicolaus of Lyra (lived *c.* 1270–1349).
(3) Moral and additional explanations.[1]

The purpose of this arrangement is proudly announced in the preface: The reader need not turn a page in order to study the commentaries, which are printed in closest proximity to the Bible text. The reader is thus supposed to understand the text in accordance with the tradition which encloses, like a large frame, the official Latin version of the Bible. It is obvious that the commentaries often take much more space than the passages explained by them, especially as the same interpretation is sometimes repeated in the different explanations. To complete the description of this Bible, which consists of six stately volumes, it must be added that references to parallel passages of the Bible are given in the margin (as they are still in the English Authorized Version).

The contents of the commentaries in all their variety and richness, offering as they do a cross-section of age-old tradition, cannot be described here. But the methodical approach is the same throughout, and two examples taken from Luke i. 26–7 may be used to signify both the importance of the actual wording of the Latin version and, resulting from it, its influence on the translator.

Luke i. 26–7 reads: 'In mense autem sexto, missus est Angelus Gabriel a Deo...ad Virginem desponsatam viro, cui nomen erat Ioseph....' The English Authorized Version has the following text: 'And (Revised Version "now") in the sixth month the

---

part 1, vol. II, no. 7763; part 2 (by F. Isaac, 1938), no. 14139. The significance of this method of exegesis is discussed by H. de Lubac, S. J., 'Sur un vieux distique: la doctrine du "quadruple sens", *Mélanges offerts au R.P.F. Cavallera* (Bibliothèque de l'Institut Catholique, Toulouse, 1948), pp. 347–66.

[1] In the first edition the moral and additional explanations are printed at the end of each chapter, while in the second edition they are found at the same page as the text.

Angel Gabriel was sent from God...to a virgin espoused (R.V.: "betrothed") to a man whose name was Joseph....' The commentaries to the words 'ad Virginem desponsatam' ('to a virgin espoused') illuminate the method used in this Bible. Nicolaus of Lyra discusses the interpretation given by earlier expositors and subdivides the exegesis under three headings: the words 'to a virgin espoused' have a special meaning in relation to the child, to the mother, and to ourselves. They are concerned with the child for four reasons: the child should never be reproached by infidels as being illegitimate; the genealogy could follow the usual method and mention the name of a man (Joseph); the birth by the virgin would be unknown to the devil; and, lastly, Joseph would bring the child up. The words 'to a virgin espoused' protect the mother from being stoned as an adulteress, they remove infamy from her, and, finally, Joseph would help the virgin. The significance of these words for us has five reasons:

(1) Joseph's testimony makes it more certain that Christ was born by a virgin (reference to Ambrose, *Supra Lucam*).

(2) The words of the Virgin asserting her virginity are made more credible. According to Ambrose an unmarried woman who is pregnant would try to disguise her fault with lies.

(3) These words remove the excuses of virgins who have incurred infamy.

(4) These words signify the Church which, though a virgin, is betrothed to a man.

(5) The fact that the mother of God was betrothed and a virgin reveals that in her are honoured both virginity and matrimony, which are thus defended against heretics.

It is noteworthy how these words are explained in their significance within the context, and how every possible doctrinal meaning is deduced from them, and how the different senses, historical, moral and allegorical, are brought to light.

The ordinary gloss, which is printed side by side with Nicolaus of Lyra, contains one addition which is interesting in its allegorical and anagogical meanings. It is taken from Ambrose and reads:

THE TRADITIONAL VIEW

'Mary espoused, that means a virgin: this typifies the Church which, though married, is immaculate....The fact that Mary married Joseph but conceived from the Holy Spirit probably means that the Church is filled by the Spirit, notwithstanding the change of the temporal priest.'

In spite of the fact that only part of the commentary on two words has been expounded here, one may be allowed to draw some conclusions on the method of interpretation, which is, as mentioned above, the same throughout the Bible. Every word is explained as having a specific significance within the text. But at the same time the special religious connotation that this word may possess, independently of the context, is used to enrich the meaning of the verse. Whatever the origin of the interpretation explaining the Church as a virgin and Christ's bride, this explanation could be applied whenever the words 'virgin', 'bride', 'betrothed', etc., are found. The reason why the allegorical explanation of Luke i. 27 (no. 4 above) can be derived from the text is very simple: both the words 'virgin espoused' carry with them this special definite religious connotation. The connexion between the actual wording of the Bible, in this case the Vulgate, and the tenets of the Church is, as can be seen from this example, very close and the Church had therefore to protect the sacred text of the Vulgate from any changes.[1]

The implications of this method for the translator are manifold. His task is to preserve the meaning of the context of the Bible as well as the meaning derived from the text by interpretation. This means that in his rendering he is to follow the authentic version of his time in all its aspects of exegesis: he should imitate the wording of the authentic version as closely as possible, and even try to preserve the order of words which, according to St Jerome, contains a mystery transcending human understanding. He should, moreover, have at his disposal a vocabulary of

---

[1] For the relationship between Scripture and Church see J. Beumer, S. J. ('Heilige Schrift und kirchliche Lehrautorität', *Scholastik*, vol. xxv (1950), pp. 50–4), who gives a survey of the views held by different Schoolmen.

religious terminology which enables him always to use the same word for rendering any specific term, for only in this case can his translation be the basis for the same exegesis as the authentic version. The imitation of the wording of the original leads to a word-for-word translation in which scanty regard, if any, is paid to the rules of the language into which the rendering is made. Medieval Bible translators up to, and including, the fifteenth century follow, generally speaking, this method and no blame should be attached to them for doing so. Their usage of words and the construction of their sentences may not always correspond to the rules of the native language. They may well have been aware of this. Yet it can safely be assumed that they wished to remain within the tradition of Bible translation, and that they purposely created a word-for-word translation without intending to substitute for the idiomatic expressions and constructions of the Latin Vulgate those commonly used in any vernacular. For the method of word-for-word translation was considered to be the surest safeguard against any alteration of the original thought. It was considered to render the contents of the Bible in its entirety without any mistake, and to protect the translator from a change of God's word and from heresy.

How easy it was to contradict the tradition may be seen from the wording of Luke i. 28, where the Vulgate reads: 'Ave gratia plena'. These words were rendered by Wyclif: 'Heil, ful of grace' whilst the Authorized Version reads: 'Hail, *thou that art* highly favoured' with a note: 'Or, *graciously accepted*, or *much graced*'. The ordinary gloss to these words is very short, stressing the significance and appropriateness of the two words 'gratia' ('grace') and 'plena' ('full'). The meaning of 'grace' is analysed: according to Gregory the Great, Mary brought forth to the world the source of the whole fullness of *grace*. Nicolaus of Lyra, quoting St Jerome, especially comments on the apposite use of the word 'plena' which connotes the fullness of grace infused into Mary. The words of the Vulgate imply the theory of infusion of grace. The term 'grace' opened therefore a way to the teacher for the full explanation of the theory of grace which, in all its aspects, had been worked out by the Schoolmen.

The importance of the Schoolmen for biblical studies can hardly be overestimated. For centuries they had erected a definite structure which they considered to be in accordance with the method of Bible interpretation and which regulated every detail of thought. However great the differences between the various schools of scholasticism, all of them considered their ideas to be in conformity with Holy Scripture, for their fundamental conception was that any view not in concord with the Bible was of necessity false. Everything therefore depended on the correct exegesis of the Bible to which the holy Fathers of the Church in their divine inspiration had shown the way.

However one-sided this view of the history of Bible interpretation may be, it was the one held by Gabriel Biel,[1] who has been called the last of the Schoolmen. He died at the end of the fifteenth century. In a remarkable preface to a compendium for students of theology he points out that just as Peter Lombard culled his *Sententiae* from the work of the Fathers of the Church to facilitate study, so his own intention is to 'abbreviate the four books of *Sentences* by the venerable *inceptor*, the Englishman William of Occam, the most piercing investigator of truth'. Biel, that modest and amiable philosopher, did not live to finish his work. His editor and successor Wendelin Steinbach reports that Biel thought himself unworthy of having his name on the title-page, 'for, as he said, he had added nothing or very little of his own to the writings of earlier philosophers'. This sentence illustrates what has been said concerning the strength and the weakness of scholasticism at the end of the fifteenth century. Its strength lay in its old tradition and a method worked out with the greatest logical subtlety. Its weakness was that the very strength of the tradition tended to inhibit the new thought necessary to the maintenance of its impetus.

[1] For Biel see J. Haller, *Die Anfänge der Universität Tübingen*, 1477–1537 (Stuttgart, 1927), pp. 153–72.

THE TRADITIONAL VIEW

The thought of the Schoolmen was, as mentioned above, closely connected with the exegesis of the Bible. Their works were studied by every student of theology who had to devote years of study to Peter Lombard's *Sententiae*. Thus any rejection of scholasticism would have the greatest repercussions upon the course of theological studies. The scholastic method was the basis of Bible exegesis and therefore the Schoolmen had to resist any change of the text of the Latin Bible, for on this very wording their philosophical thought was based.

The words 'plena gratia' of Luke i. 28 may again serve as an example. The meaning of 'gratia' in its manifold theological connotations may be omitted here and only some of the intricacies of the word 'plena' and 'plenitudo' may be mentioned. Biel distinguishes four kinds of fullness of which the first, 'fullness of sufficiency' ('plenitudo sufficientiae'), has four subdivisions: it suffices either for the acquisition of eternal life, from which fullness every mortal being is excluded, or it suffices for a task to which man is ordained, or thirdly for the grace to which man is preordained from God, or fourthly for having the nimbus or the aureole. Here 'nimbus' means the essential reward while 'aureole' is accidental only since it adds nothing essential to the nimbus. The three first kinds of the fullness of sufficiency belong to all the saints, but the last only to St Stephen, of whom it is said: 'And Stephen, full of faith and power' (Acts vi. 8). The second kind of fullness is called 'fullness of prerogative excellence' ('plenitudo excellentis praerogativae'), for he who possesses it, excels all creation not united with the Word. All sinful mortals are excluded from it. It is the fullness belonging to the 'glorious virgin' of whom it is said: 'Ave gratia plena'. The third is the 'fullness of abundance' ('plenitudo generalis copiae'). This includes every kind of grace and therefore the Church is full of grace since she possesses them all. For it is said in 1 Cor. xii. 4–11:[1] 'Now there

---

[1] The wording given is that of the A.V. except for 'gifts' which has been altered to 'graces' since the Vulgate, which is quoted by Biel, reads 'divisiones vero gratiarum sunt'.

are diversities of graces, but the same spirit.... For to one is given by the spirit the word of wisdom, to another the word of knowledge, by the same spirit. To another faith, by the same spirit: to another the gifts of healing, by the same spirit: to another the working of miracles, to another prophecy, to another discerning of spirits, to another *divers* kinds of tongues, to another the interpretation of tongues. But all these worketh that one and the selfsame spirit, dividing to every man severally as he will.' The fourth and last kind of fullness is the 'fullness of overflowing and superabundance' ('plenitudo effluentis superabundantiae'). This is the highest grace which only Christ possesses, of whom it is said: 'And we beheld his glory, the glory as of the only begotten of the Father, full of grace and truth' (John i. 14).[1]

The subdivisions of 'fullness' are used for the interpretation of the Bible wherever the word 'full' occurs. If an attempt were made to modify or change these definitions, the obvious answer would be that this structure has been built up from the Bible and without this structure the Bible cannot be sufficiently interpreted.

This way of reasoning was also put to use in sermons. Although there are some sermons which are free, or comparatively free, from the teaching of the Schoolmen, most of them are based on definitions and distinctions. It is therefore only to be expected that when speaking about the annunciation Biel should make the subdivisions of 'fullness' the subject of his sermon. In it all the subtleties of the logical method are employed for the exegesis of the Bible and every apparent contradiction in the wording of Holy Scripture is revealed and bridged over. For example, it is pointed out that the words 'plena gratia' are said before the words 'thou shalt receive' (Luke i. 31). This raises a question: The virgin was *full* of grace; logically speaking, nothing can be added to fullness; how then can the logical sequence of 'full of grace' (Luke i. 28) and 'thou shalt receive' (Luke i. 31) be explained? One of the possible solutions of this problem is, according to Biel,

---

[1] Gabriel Biel, *Collectorium in quattuor libros sententiarum* (Jacobus Wolff de Pforzheim, Basel, 1512), liber iii, Dist. 13, Quaestio unica, Art. 3 Dubi. 1, fol. F ij^r, col. a.

that 'full' means perfection which opens a new vista of differentiations of the word 'full'.[1]

The importance of the two words 'full' and 'grace' for the understanding of the Bible text forces the translator to make a literal rendering if he wishes to remain within the traditional interpretation. Wyclif's 'Heil, ful of grace' could be used as well as the Latin text to deduce the orthodox exegesis of this passage. Those, however, who in the sixteenth century broke away from the traditional view chose in the Geneva edition of 1557 to translate: 'Hayle thou that art freely beloved'. Naturally, this wording could not be reprinted in the Catholic Rheims Bible of 1582, where the literal version of the Latin text is retained: 'Haile ful of grace'. These differences in the renderings of these few words indicate the close connexion between the traditional interpretation and the translation to be chosen. They prove, moreover, that without a change of exegesis, a change in translation is unlikely. This means that every translation had to follow the Latin text as closely as possible as long as the alliance between theology and scholastic philosophy was unshaken. A breakdown of this philosophical method would entail a new aspect of exegesis and thus a new translation. It is easy for the student looking back at the fifteenth century to see that any weakening of scholasticism would imperil the interpretation of the Bible and thus weaken the Church itself, but the people of the fifteenth century were, of course, not aware of this danger.

At the end of the fifteenth century the belief in scholasticism was decreasing. Some of the Schoolmen themselves seem to have had the feeling that the end of this philosophical method was drawing near. There was Paulus Scriptor, who in 1498 published a commentary on the first book of Duns Scotus' *Sententiae*.[2] Yet he is reported to have said that theology should be changed and

---

[1] G. Biel, *Sermones de festivitatibus Christi*, the second part of which has the title *De festivitatibus gloriosae virginis Mariae* (Johann Otmar, Tübingen, 18 November 1499, 10 March 1500). The sermon referred to is Part 2, no. 21 or no. 4, delivered at the feast of Annunciation, fol. K6v, col. b to L3v, col. a.

[2] Printed by Otmar (Tübingen, 1498).

that scholasticism should be replaced by the study of the Fathers of the Church.¹ There was Wessel Gansfort who is said to have held similar views. Wessel died in 1489, but Luther claimed him as a predecessor. He was a remarkable man who had visited Greece and Egypt and who knew Hebrew and Greek. He was known as 'lux mundi' and 'master of contradiction',² for his views differed in many respects from those of his contemporaries. For this reason he was reproached in a letter with these words: 'And unquestionably, in view of being a most learned man, your singularity gives offence to many.'³ Yet, whatever his theological views he considered logic as an essential part of theology.

For who [he wrote] could ever attain to that apex of theology, to which Peter D'Ailly climbs, without definitions, divisions, argumentations, distinctions, and logical instances? I am speaking of disputations, where there is need of the sharp tooth of discussion; not of sermons to the people, nor of meditation Godward.... Theologians must have recourse to logic.⁴

The same distinction between logic and religious feeling is made in a letter to a nun who had asked him about the study of logic. The passage in which he expresses his view that women cannot think logically and yet reach the highest aim through love, is so personal and full of charm that I will quote it in full:

With regard to the study of logic, I do not deny that it contributes to scholastic discipline. But I do not see what it adds to the consolation of monastic solitude and spiritual exaltation, especially in the case of maidens like yourself. As a rule it has been given to your

---

¹ Mentioned by Konrad Pellikan, *Chronikon*, ed. B. Riggenbach, p. 24.
² Reported by Gerard Geldenhouwer, printed in J. Fichardus, *Virorum qui superiori nostroque saeculo eruditione et doctrina illustres atque memorabiles fuerunt, Vitae* (Frankfurt, 1536), pp. 87ᵛ–88. Wessel, *Farrago, Epistolae, Tractatus de Oratione*, first printed (without place and date), *c.* 1521, fol. b iiijʳ. Literature on Wessel: E. W. Miller, *Wessel Gansfort* (I generally follow the translations printed in this book; they are by J. W. Scudder), 2 vols. (New York and London, 1917); P. S. Allen, *The Age of Erasmus* (Oxford, 1914), pp. 9–13, 29–32; M. van Rhijn, *Studiën over Wessel Gansfort en zijn tijd* (Utrecht, 1933).
³ *Ep.* from Jacobus Hoeck of unknown date, printed, *op. cit.* fol. b iiijʳ, translated by Scudder, *loc. cit.* vol. I, p. 277.
⁴ *Ep. De Indulgentiis*, to Hoeck, of unknown date, ch. 9, fol. e iiijʳ, translated by Scudder, *loc. cit.* vol. I, pp. 308–9.

entire sex to glow with eager longing rather than to be distinguished for judgment or discernment. Hence I think the highest logic for you consists in prayer. For the promise, 'Seek and ye shall find', has not been given in vain to you. Long before you could learn logic, you will have prevailed through the prayer of faith with the Teacher of truth to grant you all needful truth. It is not expedient that the eye of the guileless dove be confused by too many things. They that too curiously consider the things that stand about their pathway, press on to the end the more slowly. Acquire love through prayer, and you will have obtained all the fruit of logic, of knowledge and wisdom.[1]

One sees Wessel, good-natured, smiling, while he is writing these words after having, in the same letter, calmed the nun about a ghost of which there was much talk among the people.

Words like these were no real danger to the traditional view. Wessel's opinions, shared by the Brethren of the Common Life, could well be maintained without a rejection of scholasticism. They were not strong enough to break a philosophical system. They appealed to the emotional capacities of man, but could neither prove that the Schoolmen were wrong nor replace their philosophy. In fact, Wessel's words show that he did not wish for the destruction of the traditional philosophy.

Most students of fifteenth-century thought would call Wessel a humanist. It is necessary to remind the reader that any humanist who at that time wished to write a book on philosophy was bound to follow the scholastic method and to use the scholastic terminology, for no other method or terminology existed. Rudolph Agricola, the greatest of the early humanists north of the Alps, who died in 1485, wrote his *De Inventione Dialectica* in a form which was scholastic. Even Erasmus, when forced to write against Luther on free will (*De Libero Arbitrio*) in 1524, followed this tradition. This can only be understood if the usual conception of an opposition between humanism and scholasticism is discarded. The teaching of the early humanists was in no way concerned with logic and scholastic philosophy. Their interests lay on a different

---

[1] Wessel, *op. cit.* fol. h^{r-v}, translated by Scudder, *loc. cit.* vol. I, pp. 241–2, with a notable difference in the last sentence.

plane. They proclaimed that style, clarity and presentation of a subject were an essential part of literature. These new conceptions were destined to undermine the scholastic method and come into conflict with it, but this attitude did not originally aim at altering the foundations of faith. All the early humanists were good churchmen, even Wessel who was one of the few who, with a knowledge of Hebrew and Greek, might have been able to challenge the traditional method of exegesis. It was possible to be a humanist and at the same time to praise the Schoolmen. Henricus Bebel, for instance, wrote verses eulogizing Gabriel Biel, which he published at the beginning of Biel's Compendium on Occam's Sententiae. The Bible edition of 1498, quoted above for the elucidation of the traditional interpretation, was edited by a humanist who had also published Gratian's *Decreta* in 1493. This was Sebastian Brant, the author of the *Ship of Fools*, first published in German in 1494 and soon to be translated into Latin, French and English.

Brant's attitude towards the Bible is the same as that of any humanist of the fifteenth century. He rejects any exegesis of the Bible that does not conform to the doctrines of the Church. He derides people who believe that they are able to explain Holy Scripture in the light of their individual reason. In this they mostly make mistakes, he says. These interpreters are false prophets whose explanations are different from those of the Holy Ghost and from ecclesiastical tradition. They distort the meaning of Holy Scripture.[1]

It is important to see that Brant, a humanist, maintains that the personal view of any individual must needs be wrong if it conflicts with the orthodox view. Many of the early humanists criticized scholasticism, and openly condemned the abuses of the Church, but although they wished for some religious reforms, they re-

---

[1] See esp. no. 103, called 'Vom endkrist' in Brant's *Das Narrenschiff*, ed. F. Zarncke (1854). Two English editions, translations from the Latin of Locher, were made by Henry Watson, printed by Wynkyn de Worde in 1509, and by Alexander Barclay, published by Pynson in 1509. These English paraphrases are very long-winded, even when compared with the original German, which is by no means terse.

THE TRADITIONAL VIEW

mained within the orbit of the Church and their challenge was not thought to be a danger to the 'traditional view'.

Even in the sixteenth century, after Luther's break with the Catholic Church, many humanists still did not believe that their fight against the Schoolmen impaired the strength of the Church. It has been shown in the first chapter that Sir Thomas More was opposed to all modern Bible translations because the expressions used in them were opposed to the tradition of the Church. Yet he defended Erasmus's work on the Bible. His view is very similar to that of Sebastian Brant: the Church alone can decide whether an interpretation of the Bible is correct. The personal view of an expositor is wrong and heretical if it is opposed to the tradition of the Church. His words taken from his *Dialogue Concerning Tyndale* of 1528 are:

...And now since ye grant, and I also, that the church cannot misunderstand the scripture to the hindrance of the right faith in things of necessity [to salvation], and that ye also acknowledge this matter to be such, that it must either be the right belief and acceptable service to God or else a wrong and erroneous opinion and plain idolatry, it followeth of necessity that the church doth not misunderstand those texts that ye or any other can allege and bring forth for that purpose.... For I have known, quod I, right good wits, that hath set all other learning aside, partly for sloth refusing the labour and pain to be sustained in that learning, partly for pride by which they could not endure that redargucion [=refutation] that should sometime fall to their part in dispysicions [=disputations].... And because they have therein the old holy doctors against them, they fall to the contempt and dispraise of them, either preferring their own fond [=foolish] glosses against the old cunning [=wise, learned] and blessed fathers' interpretations, or else lean to some words of holy scripture, that seem to say for them against many more texts that plainly make against them, without receiving or eargiving to any reason or authority of any man quick or dead, or of the whole church of Christ to the contrary. And thus once proudly persuaded a wrong way, they take the bridle in their teeth, and run forth like an headstrong horse that all the world cannot pluck them back. But with sowing sedition, setting forth of errors and heresies, and spicing their preaching with rebuking of priesthood and prelacy for the peoples' pleasure, they turn many a man to ruin and themself also. And then the devil deceiveth them in their blind affections.[1]

Thomas More's opinion is in concord with the opposition of the Church to an exegesis based on personal thought. The Council of Trent was to reaffirm this in the following decree:

Praeterea ad coercenda petulantia ingenia decernit, ut nemo suae prudentiae innixus, in rebus fidei et morum, ad aedificationem doctrinae Christianae pertinentium, sacram Scripturam ad suos sensus contorquens, contra eum sensum quem tenuit et tenet sancta mater ecclesia, cuius est iudicare de vero sensu et interpretatione Scripturarum sacrarum, aut etiam contra unanimem consensum Patrum ipsam Scripturam sacram interpretari audeat, etiamsi huiusmodi interpretationes nullo unquam tempore in lucem edendae forent. Qui contravenerint per ordinarios declarentur, et poenis a iure statutis puniantur.[2]

[1] Thomas More, *The English Works*, ed. W. E. Campbell and A. W. Reed, vol. II, pp. 78–80.
[2] 'Decretum de editione et usu sacrorum librorum', *Canones et Decreta Concilii Tridentini*, ed. A. L. Richter (Leipzig, 1853), p. 12.

CHAPTER IV

THE PHILOLOGICAL VIEW: REUCHLIN

The strength of the Church lay in the immense power of the centuries-old tradition against which views of individuals meant little. Who, after all, had the right to condemn this tradition and thus to imply that God had misled Christianity for so long? If someone wished to change Bible interpretation, he could not rely on his own personal views, but had to use a new method which in its turn was to replace the old one. A method is more impersonal than subjective reasoning. A new method can perhaps reveal errors in the old one and thus open the way to a new approach based on method and not on the thoughts or whims of any one person.

But there existed no method with which to fight the combined forces of theology and scholasticism. The early humanists, as mentioned above, were not concerned with theology and philosophy. By the end of the fifteenth century, however, the humanists north of the Alps had clarified their aims and, partly in self-defence and partly in pride of their own accomplishments, they proclaimed the value of their studies and contrasted them with scholasticism. Their arguments, which were borrowed from Italian humanists, especially from Laurentius Valla, were based on the belief in the value of language. A civilized language, they thought, created a flourishing civilization. Any deterioration in the standard of language entails a decay in civilization. From this it was concluded that a return to the Latin style of antiquity would be followed by a new, golden age of learning and civilization. The first task of the humanist was therefore to read the Latin authors of antiquity. The early Fathers of the Church and the early Christian poets were included among those whose style was considered to be praiseworthy. But the Schoolmen were excluded, and the reading of their volumes was thought to be useless. It was therefore logical for the humanists to edit the texts of the authors of antiquity, of the old Christian writers, and of the Fathers of the Latin Church.

In the literary field the humanistic conception of the importance of language made it desirable to study the originals rather than translations, for it was recognized that the actual wording used by the author was an essential part of a literary work. Presently the study of language was extended from Latin to Greek and Hebrew, thus arousing interest in Greek poetry and philosophy. It was of the utmost importance when, in accordance with this tendency, the authoritative Latin text of the Bible was laid aside and Holy Scripture was re-interpreted from Hebrew and Greek texts. No attempt was made to replace the Vulgate in the services of the Church, though new translations were made in considerable numbers. The humanists were able to show that some of the difficulties of the Vulgate text were due to ambiguities or even mistakes in the Latin translation. The principle, as expressed by Laurentius Valla in his *in Latinam Novi Testamenti Interpretationem . . . Adnotationes*,[1] was clear. Nobody was entitled to interpret the sacred text unless he was able to read it in the original Hebrew or Greek. Thus a new reason was adduced against the exegesis of the Schoolmen, who knew Latin only and based their work on the Vulgate alone. The humanists were able to refer to those Fathers who had known Hebrew and Greek and to rely on their explanations rather than on those of the later commentators.

It has been said that the battle-cry of sixteenth-century humanism was: '*Ad fontes*', 'Back to the sources!'.[2] This methodical return to the sources had two important results: by drawing attention to the original texts it was possible to go back to a tradition earlier than that of the Schoolmen and thus to adduce testimonies recognized by everybody. Moreover, stress could be laid on the great importance of the early Christian writers. This would justify the complete neglect of the Schoolmen. The second result was the recognition that every interpretation has to start

---

[1] Edited by Erasmus (Badius, Paris, 13 April 1505).
[2] J. Huizinga, *Erasmus of Rotterdam*, translated by F. Hopman (London, 1952), p. 109.

from the original and not from the translation. This assumption, which is still valid to-day and which is one of the lasting achievements of humanism, obviously led to a lowering of the value of the Vulgate. The prestige of the Vulgate was further impaired by the new translations in which the traditional text was considerably changed in accordance with the original Hebrew or Greek. It was at this stage that defenders of the 'traditional view' joined issue with the humanists. This fight greatly weakened their influence and thus diminished the authority of the Vulgate still further. The ascendancy of humanism is therefore an important factor in the preparation of the ground for new Bible translations. It is for this reason that consideration must be given here to the development of humanistic thought towards a new method of Bible interpretation and translation.

This development was achieved mainly by two men, Reuchlin and Erasmus, whose Bible studies in Hebrew and Greek respectively occupied a central place in their life's work. Therefore their attitude towards Bible studies will be placed in the framework of their thought, indicating the difficulties they had to overcome to reach the aim for which they fought.

It is difficult to arrive at a full understanding of the obstacles to Hebrew studies in the fifteenth century. Why should they be pursued at all? Their application to the sacred texts and thus to Theology might seem to us a sufficient reason. Yet in the fifteenth century this was not acceptable. Peter Schwarz (Nigri), a Dominican, could assert that the occupation with Hebrew would turn people away from the study of 'vain profane letters'.[1] The use of Hebrew for the reconstitution of the Vulgate was very doubtful. The text of the Vulgate was in a bad state, but there was no certainty that the Hebrew wording was not corrupt. On the contrary, it could be safely assumed that the Jews had purposely

---

[1] Peter Schwarz (Nigri) *Super Psalmos*, Prologue written in 1476/7, published by B. Walde, 'Christliche Hebraisten Deutschlands am Ausgang des Mittelalters', *Alttestamentliche Abhandlungen*, vol. VI, parts 2–3 (1916), pp. 89 ff., esp. p. 91.

falsified the text, at least in all those sentences where Christ was announced.[1] The task of the Christian obviously was to teach the Jews their wilful perversion of the Bible and in this way to induce them to become Christians. The other way, the correction of the Latin text, in accordance with the Hebrew, was dangerous since it implied that the Christian Bible was faulty. Jewish interpretation of the sacred text was thought to have been written to mislead the reader. Therefore these works should be read by nobody. This view is clearly expressed by Peter Schwarz (Nigri) in his *Stella Meschiah* of 1477. In it he sets out the reasons for the publication of this book, some of which are: Christians should be aware of, and guard themselves against, the false interpretations of the Jews who reject those Biblical sayings which refer to the Christian faith; the *Stella Meschiah* will furnish them with the answers to Jewish attacks upon the Christian faith; the book will advise Jews to become Christians.[2]

Talmud and Cabbala were thought to be filled with falsehoods hindering the true understanding of the Bible and attacking Christianity. Peter Schwarz adduces these reasons against the Talmud,[3] whilst Pico della Mirandola in his *Apologia* of 1487 obviously speaks against the accepted view when he points out that the Cabbala cannot have been falsified by the Jews since it was written down before Christ. The Jews had therefore no reason to corrupt the truth of the Cabbala, which had been given to them by God. In defence of his studies Pico explains that the Cabbala almost completely agrees with Christian teaching and can be turned against the Jews who, holding its teaching in great esteem, will be unable to deny its truth.[4] The belief in the falsification of the Hebrew writing by the Jews hindered the study of Hebrew, for why should these falsehoods be learned? A complete reversal of the accepted opinion was necessary before the study of Hebrew

---

[1] This was the view of, for example, Nicolaus of Lyra (see *P.L.* vol. CXIII, cols. 29–30).
[2] *Stella Meschiah* (Conrad Feyner, Esslingen, 20 December 1477) (without fol. and pag.), leaf 10^{r-v}. [3] *Stella Meschiah*, leaf 286^r.
[4] *Apologia*, chapter containing 'De Magia naturali et Cabala', *Opera* printed by Sebastian Henric-Petri (Basel, 1601), vol. I, p. 116.

could seriously be started. When Reuchlin attempted to use the Hebrew texts and rabbinical explanations for the elucidation of Scripture, he had to withstand the fury of those who despised Jewish tradition. He had to fight against them all through his life and at the end, tired, disappointed and bitter, he wrote of himself that he was, 'as it were, the protomartyr of Hebrew letters, sacrificed, tortured, burned, torn to pieces and lacerated'.[1]

Another difficulty for the student of Hebrew in the fifteenth century was the lack of good teachers. The Jews had been expelled or were unable or unwilling to teach Hebrew. Pellican writes in his autobiography that he could find no Jew who was able to explain any grammatical question.[2] Reuchlin in his Hebrew Grammar and Dictionary, called *De Rudimentis Hebraicis*, mentions that the Jews, invoking the authority of the Talmud, refused to teach Hebrew: 'The Jews of our country', he says, 'wish to instruct no Christian in their language because of ill-will or ignorance.'[3] Whatever the reason for this reticence may have been, it is a fact that it was very difficult or even impossible to find a teacher of Hebrew. There were baptized Jews. One of them, Johannes Pauli,[4] taught Pellican, but it is very doubtful whether his knowledge was sufficient for this purpose. Pellican's grasp of Hebrew after this teaching was not even elementary, since he had not yet understood the basis of Hebrew verb formation.[5] Reuchlin learned from several Jews, especially from the Emperor's physician, Jacob ben Jehiel Loans, who became his teacher in 1492.[6] Besides, he made extensive use of the works of Hebrew grammarians.

[1] *R.E.* no. 303, p. 333.
[2] *Das Chronikon des Konrad Pellikan*, ed. B. Riggenbach (Basel, 1877), pp. 19–20. Cf. *R.E.* no. 256, p. 296. The bibliography of the development of Hebrew studies is given by O. Kluge, 'Die hebräische Sprachwissenschaft in Deutschland im Zeitalter des Humanismus', *Zeitschrift für die Geschichte der Juden in Deutschland*, vol. III (1931), pp. 81–97, 180–93. A very short survey is found in D. Daiches: *The King James Version of the English Bible* (Chicago, Illinois, 1941), pp. 128–38.
[3] *De Rudimentis Hebraicis* (Thomas Anshelm, Pforzheim, 27 March 1506), p. 621.
[4] Pellican, *Chronikon*, pp. 14, 16. Pauli was the author of *Schimpf und Ernst*.
[5] Pellican, *op. cit.* p. 19.
[6] *De Rud. Hebr.* p. 619. Cf. L. Geiger, *Das Studium der hebräischen Sprache in Deutschland* (Breslau, 1870), pp. 18 ff.; idem, *Johannes Reuchlin* (Leipzig, 1871), pp. 105–6; J. Perles, *Beiträge zur Geschichte der hebräischen und aramäischen Studien* (München, 1884), p. 30.

THE PHILOLOGICAL VIEW: REUCHLIN

The last difficulty to be mentioned here was the scarcity of Hebrew grammars and books. As the beginner could not use grammars written in Hebrew, he had to rely on the little primer of six leaves by Peter Schwarz (Nigri) found at the beginning of his *Stella Meschiah*. This is an instruction in German on how to read Hebrew. But neither this book nor his *Tractatus contra Judaeos*[1] of 1475 contained long passages printed in Hebrew characters, but only transcriptions together with interlinear translations into Latin or German. In about 1501 Aldus Manutius published a short primer of Hebrew on eight leaves called *Introductio Utilissima Hebraice Discere Cupientibus*[2] which was frequently reprinted as *Introductio Perbrevis ad Hebraicam Linguam*, e.g. by Nicolaus Marschalk at Erfurt in 1501 or 1502, and later in various editions in C. Lascaris's *De Octo Partibus Orationis*.[3]

The first grammar ever to be published in a European language was by Conrad Pellican, who later was to become a follower of Zwingli. It was published in 1503 or 1504, and was poor in content with types neither beautiful nor clear. Pellican had gleaned his knowledge from various sources. In addition to his early instruction from Pauli, he had been in contact with Reuchlin and had read Peter Schwarz's books and probably a Hebrew grammar which he had found in the library of the priest John Behaim at Culm. This grammar had been translated into German by a Jew who was ignorant of Hebrew grammar.[4]

In 1506 Reuchlin's *De Rudimentis Hebraicis* was published,

---

[1] Conrad Fijner, Esslingen, 6 June 1475, sine fol. et pag. This is the first book printed in Germany in which Hebrew characters (woodcuts) are found, three Hebrew words on leaf 10.

[2] For the authorship of Matthaeus Adrianus see G. Bauch, 'Die Einführung des Hebräischen in Wittenberg', *Monatsschrift für die Geschichte des Judentums*, N.F. 12 (1904), pp. 332 and *E.E.* vol. III, no. 686 (5–11), esp. note 5.

[3] For bibliographical details see A. Marx, 'Some Notes on the Use of Hebrew Type in Non-Hebrew Books, 1475–1520', in *Bibliographical Essays, A Tribute to Wilberforce Eames* (Cambridge, Mass., 1924), pp. 386 ff.

[4] Pellican, *Chronikon*, p. 19. Pellican's *Grammatica Hebraea* was printed in G. Reisch, *Aepitoma omnis phylosophiae, Alias Margarita Phylosophica...tractans de omni genere scibili cum additionibus quae in aliis non habentur* (Johannes Grüninger, Strassburg, 1504). The Grammar is on fol. F ix^r–xxviij^r. The date of the first edition is probably 1503, cf. A. Freimann, *A Gazetteer of Hebrew Printing* (New York, 1946), s.v. 'Basel', p. 17.

THE PHILOLOGICAL VIEW: REUCHLIN

marking the beginning of Hebrew studies in Europe. From this date on it was possible to learn Hebrew.

Hebrew manuscripts and books were rare north of the Alps. It is true, there were some libraries with Hebrew manuscripts such as that in the Dominican Monastery at Regensburg which received forty-three manuscripts in 1476, but there was no Old Testament among them.[1] There were several libraries, such as Behaim's library which was used by Pellican, and the famous library of Cues. Printed books were published by and for the Jews in Italy, but only few found their way to Germany. The first print was Rashi's Commentary on the Pentateuch published at Reggio Calabria on 17 February 1475, and the first complete Old Testament (folio) was published by the Soncino family at Soncino in 1488, the first octavo edition being printed at Brescia in 1494. The number of Hebrew Bibles is considerable.[2] Conrad Pellican gives a vivid description of his endeavours to obtain manuscripts or books. He tells how a magnificent manuscript of parchment with three columns on each page, containing Isaiah, Ezechiel, and the twelve minor prophets, was carried to him. He describes a visit to the library of John Behaim without, however, giving details about the books in it. In 1500 he at last received a complete Hebrew Bible[3] which had been printed in Italy.

---

[1] The library at Regensburg, see B. Walde, *op. cit.* pp. 74–82; Behaim's library, see *ibid.* pp. 190–9.

[2] A description of Hebrew Bibles printed before 1500 is found in *Gesamtkatalog der Wiegendrucke* (Leipzig, 1925–), s.v. 'Biblia', nos. 4198–4200. A list of Hebrew incunabula is compiled in *The Jewish Encyclopedia*, new ed. (New York and London), s.v. 'Incunabula', vol. VI, pp. 578–9; vol. IX, pp. 463–4, s.v. 'Soncino', where the publications of the Soncino family are enumerated. For Hebrew printing see A. Freimann, *op. cit.* and A. Marx, *op. cit.* (Many of the bibliographical details of Hebrew printing in L. Geiger, *Johann Reuchlin*, are obsolete and need correcting.)

[3] Pellican, *Chronikon*, pp. 16–20. Pellican wrote this autobiography in 1544, forty-four years after he received the Bible. He makes a mistake when saying that the Bible was printed in Pesaro where no Bibles were printed before 1511. As he speaks of a Bible of very small size ('minima forma') (p. 20), he can refer to the octavo edition of 1494 only. Another probable slip of memory, mentioned by B. Walde, *op. cit.* p. 86 n. 2, is found in his reference to Peter Schwarz's (Nigri) book *Stella Meschiah*, which has a German interlinear translation, not a Latin one as Pellican asserts (p. 17). He seems to refer to Schwarz's *Tractatus contra Judaeos* where his quotation is found (leaf 28^v). A MS. containing a Latin translation of *Stella Meschiah* of the beginning of the sixteenth century is mentioned by Walde, *loc. cit.* p. 86 n. 2.

THE PHILOLOGICAL VIEW: REUCHLIN

Reuchlin's experience in this respect is similar. We learn that in 1491 and 1492 no Hebrew Bibles could be bought at Florence. In 1506 he wrote that Hebrew books were printed everywhere. Yet new difficulties arose because wars made the import of books into Germany difficult. Soon Reuchlin found out that his Hebrew Grammar and Dictionary was of no great use so long as there were no Hebrew books for the student. Therefore in 1512 he published *In Septem Psalmos Poenitentiales Hebraicos Interpretatio*, the first Hebrew text to be printed in Germany.[1] Later, in spite of some difficulties the position became easier and books became gradually available.[2]

What were the reasons for the awakening of interest in Hebrew? A separate answer must be given for each one of the humanists who studied this language. It is an important question since it leads to the clarification of thought in the fifteenth and sixteenth centuries. It provides the key for the understanding of those whose work was to undermine the position of the Vulgate.

Peter Schwarz's position is not clear. He learned Hebrew with Jewish children when in Spain. He even compared at least the

---

[1] See Preface, fol. a vi^v.

[2] The following bibliographical notes containing information about Hebrew books may be of interest: *R.E.* nos. 31 and 33, pp. 31, 32 of 1491 and 1492. The texts of these letters are in J. Reuchlin, *Illustrium Virorum Epistolae, Hebraicae, Graecae et Latinae* (Thomas Anshelm, Hagenau, May 1519), fol. a. iv^v, and Reuchlin, *Clarorum Virorum Epistolae* (Thomas Anshelm, Tübingen, March 1514), fol. b. ii^v; *De Rudimentis Hebraicis* (1506), p. 1; *In septem Psalmos poenitentiales* (Thomas Anshelm, Tübingen, 1 August 1512), fol. a vi^v (*R.E.* no. 151, p. 175, no. 152, p. 177 of 1512). Two letters of Reuchlin's of 1520 and 1522, published by A. D. Horawitz, *Zur Biographie und Correspondenz Johannes Reuchlins*, nos. 39 and 44, *S.W.A.*, vol. LXXXV (1877), pp. 182, 187. Cf. Hummelsberger's letter to the printer Thomas Anshelm of 1512, published by Ad. H. Horawitz, 'Analecten zur Geschichte des Humanismus in Schwaben (1512–1520)', *S.W.A.*, vol. LXXXVI (1877), pp. 235–7.

Melanchthon had two Hebrew Bibles sent from Leipzig to Wittenberg, one of them commented, in 1518 (*Opera*, ed. C. G. Bretschneider, in *Corpus Reformatorum*, vol. I (Halle, 1834), no. 19, col. 43; cf. no. 22, col. 48). Boeschenstein who taught Hebrew at Wittenberg in 1518–1519 possessed the Hebrew Bible printed at Brescia in 1494. This copy is now at Trinity College, Cambridge, some missing pages are written by Boeschenstein himself who added his signature. Luther also had the Brescia edition of 1494 of the Hebrew Bible. His copy was at Berlin until 1939. The contents of Reuchlin's famous library comprising many Hebrew manuscripts and books have been listed by K. Christ, 'Die Bibliothek Reuchlins in Pforzheim', 52. *Beiheft zum Zentralblatt für Bibliothekswesen* (Leipzig, 1924), *passim*, esp. pp. 6–8, 36–51.

THE PHILOLOGICAL VIEW: REUCHLIN

Hebrew text of the Psalms with the Latin translation which, he discovered, was not made by St Jerome. However, he cannot be regarded as a humanist because, as mentioned above, he hoped that people who learn Hebrew would not read profane books. His importance for this study is mainly based on the fact that his introduction helped Pellican to learn the rudiments of Hebrew.

In 1489, when about 11 years old, Conrad Pellican[1] heard that in a disputation a certain doctor of theology had been confuted by the answers of a Jew and even of a Jewess. The fact that a Jew got the better of the argument in a religious disputation (a fact hardly ever recorded)[2] made a deep impression on the boy:

When as a boy I heard this [he writes in his autobiography fifty-five years after the event] I was greatly astonished and grieved whilst my conscience was severely tempted that our Christian faith was not supported by more solid arguments than those which could be torn to pieces by the Jews against the learned theologians.[3]

He was confirmed in these doubts when later, as a monk, he learnt 'that the mysteries of Scripture were not so clear'. At the time when he did not yet understand the divine oracles of the prophets he was 'perturbed' by the fact that explanations like those of Nicolaus of Lyra and Paul of Burgos, which were based on the Jewish expositions, were sometimes preferred to those of the Christian tradition and that 'the Hebrew truth' was quoted against the Latin Vulgate. All these experiences aroused his wish to learn Hebrew.[4]

Because of the discrepancies between Christian and Jewish interpretation he deemed it necessary to find out the true meaning of the Bible through his study of the original Hebrew text. Pellican's critical sense forbade him to condemn all non-Christian thought without examination. Even if the expression 'Hebrew

---

[1] The latest short appreciation of Pellican is given by O. Kluge, *loc. cit.* pp. 89-90. For bibliography see also Hauck, *Realencyclopaedie*, vol. xv, s.v. 'Pellican', pp. 108-11.
[2] Andrew of St Victor 'admits with cheerful indifference, that he cannot hold his own against the Jews in argument'. Mentioned by B. Smalley, *The Study of the Bible in the Middle Ages* (Oxford, 1952), p. 119.
[3] *Chronikon*, pp. 14-15.   [4] Pellican, *Chronicon*, pp. 1-15.

truth' reflects his thought when writing his autobiography and not when starting his Hebrew studies, the wish for independent investigations caused him to examine the sources of the original language.

Reuchlin, the great scholar, who at the end of his Hebrew Grammar proudly quotes Horace's 'Exegi monumentum aere perennius', was born on 22 February 1455. In a short sentence written for the preface of his second work on Hebrew language *De Accentibus et Orthographia Linguae Hebraicae* published in 1518 he offered an explanation of his life's work: 'Above all', he says, 'I have pursued the study of many foreign languages with such an exertion as well as eagerness that I have no doubts about having followed the beacon of the Genius consuming me.'[1] And a few lines later he lays emphasis on his piety in these words: 'ingenio trahor et amore pietatis'. His first publication was a Latin dictionary, called *Vocabularius Breviloquus*, of 1475 or 1476; his propagation of Greek is noteworthy, and his crowning achievement was the introduction and dissemination of Hebrew studies in Germany. He was a lawyer who was ennobled by the Emperor and who was one of the three Judges of the Swabian Confederation. His friends describe him as friendly and kindly.[2]

Unfortunately, we know of no statement by Reuchlin of the final impulse that led him to his studies. At first sight his work seems to contain incompatible features. He is a philologist, a keen observer of the connotations of words and their grammatical

[1] P. II^v. The book was published by Thomas Anshelm, Hagenau, February, 1518.
[2] *E.E.* vol. IV, no. 1006 (36–7). Bibliography on Reuchlin: *B.D.G.* nos. 17841–89. The book that has influenced almost all the later works on Reuchlin is L. Geiger, *Johann Reuchlin*, and his subsequent article in *A.D.B.*, vol. XXVIII, pp. 785–99. I do not wish to give a detailed criticism of this important book. Since my view of Reuchlin is widely different from that of Geiger, I do not intend to refer to him at every point of divergence. G. Kawerau's article in Hauck, *Realencyclopaedie*, vol. XVI, pp. 680–8 should be especially mentioned. Additional to *B.D.G.*: A review of Geiger's book by Sir Adolphus Ward, *The Saturday Review* of 25 November 1871, reprinted in *Collected Papers*, vol. III, Literary (i) (Cambridge, 1921), pp. 71–85; S. A. Hirsch, 'Johann Reuchlin, the Father of the Study of Hebrew among Christians', in *A Book of Essays* (London, 1905), pp. 116–50; W. S. Lilly, *Renaissance Types*, ch. 4, 'Reuchlin—the Savant' (London, 1901), pp. 175–230. Important detailed information is given by B. Walde, *loc. cit.* pp. 36–8, 184–6.

THE PHILOLOGICAL VIEW: REUCHLIN

peculiarities, and at the same time, being a Cabbalist, his exegesis of the contents is often an affair of philosophical and mystical explanation. Modern research has not been able to see unity in these discrepancies. Reuchlin has been highly praised as a philologist, but his philosophical views have been condemned as 'misbegotten progeny of a sick mind'[1] and 'whirling thoughts',[2] or 'it was but lost labour that he rose up early, and late took rest, and ate the bread of carefulness, in order to spin this cobweb'.[3] Yet in spite of all these remarks nobody is able to say whether Reuchlin persevered in these philological studies because of these 'aberrations', or whether philology led Reuchlin to these 'confused thoughts', for strange as it may seem, it is often difficult to disentangle these two aspects of his work.[4]

If in the following pages the philological tendencies are discussed first, it is because this part of Reuchlin's work illuminates his thought most clearly and bears most closely upon the question under discussion—what was the humanistic principle of Bible interpretation?

In 1488 Reuchlin mentions in a letter that under the influence of St Jerome he has learned Hebrew and Greek in order to read Holy Scripture in the language 'in which they are believed to have been originally composed under God's inspiration'. He continues: 'Wine that is often drawn off the cask loses in splendour. The same applies to translations: the original language of every work is sweetest.'[5] This statement on translation, carefully worded though it is, implies that Reuchlin prefers the Greek and Hebrew texts of the Bible to the Vulgate. Yet he could not be reproached for this predilection, for his words refer to Jerome's preface to the

[1] Geiger, *A.D.B. loc. cit.* p. 793.
[2] C. Wille, 'Johann Reuchlin', *Zeitschrift f. d. Geschichte des Oberrheins*, vol. LXXVI, N.F. 37 (1922), p. 263.
[3] Lilly, *loc. cit.* p. 224.
[4] G. Knod ('Findlinge', *Zeitschrift f. Kirchengeschichte*, vol. XIV (1894), pp. 118–19) believes that theological interest was the reason for Reuchlin learning Hebrew. Knod finds the proof for this in a letter written between 1482 and 1485 where Reuchlin asks Rudolf Agricola's advice on the meaning of some Hebrew words in Psalm liv. 3. It seems to me that this very short letter is no sufficient evidence for a decision either way.
[5] *R.E.* no. 15, p. 16.

THE PHILOLOGICAL VIEW: REUCHLIN

translation of the Proverbs which Reuchlin was to quote in his *In Septem Psalmos Poenitentiales* of 1512.[1] For the defence of his view Reuchlin leant on canon law, which he quoted when he advised the Emperor against the burning of Hebrew books in 1510:

> Augustine says in his *De Vera Religione* that the language of holy Scripture should be understood according to the peculiarity of every particular language, for each language has its own particular manner of speech.... When one language is translated into another, then everybody believes that it makes no sense and does not sound right. Thus it is written in canon law distinctio xxxviii. ca. locutio.[2]

In 1513 he is more outspoken, saying:

> I am doubtful about the translations which have often caused me to err in the past. Therefore I read the New Testament in Greek, the Old Testament in Hebrew, in the exposition of which I trust myself rather than anybody else.[3]

With these words he boldly asserts errors in the traditional explanations. He was confident of his ability to interpret the Bible better than anybody else. Therefore he compared the Vulgate with the Hebrew text. For him only the text of Scripture in its original language is an authority not to be doubted. An example may show what was the basis for this conviction.

At Reuchlin's time it was unknown to the Gentile students of Hebrew that the Hebrew vowel system was a creation of the Middle Ages. Reuchlin wished to prove that the Seventy translators used a Hebrew text without vowels and that some of their mistakes could be explained as resulting from misreadings of unpunctuated words. His argument is: By a change of the vowels

---

[1] St Jerome, *P.L.* vol. XXVIII, col. 1244; Reuchlin, *In Septem Psalmos Poenitentiales*, fol. a viii^r.

[2] *Augenspiegel* (Thomas Anshelm, Tübingen, 16 August 1511), p. IIII^v. Canon Law, *Decreta*, distinctio xxxviii, c. xiv is borrowed from Augustine, *De Vera Religione*, c. 50 (*P.L.* vol. XXXIV, col. 166). Reuchlin's *Augenspiegel* has been reprinted by H. von der Hardt, *Historia Literaria Reformationis* (Frankfurt and Leipzig, 1717), part II, pp. 16–53. The above sentence is p. 23 a.

[3] *R.E.* no. 163, p. 189. Cf. Erasmus, Preface to Laurentius Valla, *Adnotationes* (1505), reprinted in *E.E.* vol. I, no. 182 (188) '...malim ego meis oculis cernere quam alienis'.

THE PHILOLOGICAL VIEW: REUCHLIN

in the Hebrew text the word of Psalm cxxix. 4 (cxxx. 4) having the letters *th, wau, r, aleph* may be read either as 'thora', meaning 'law',[1] or as 'thiware' meaning 'you will be feared'. The Vulgate has 'legem', but Jerome's translation reads 'quia terribilis es' (Authorized Version: 'thou mayest be feared'). Reuchlin explains this difference by saying that the Seventy translators mistook the one Hebrew word for the other. But he goes one step further. He has noticed that some of the Greek manuscripts of the Septuagint have the reading 'ὄνομα', meaning 'name'. He tries to explain this as a chain of mistakes: The Seventy translators in their 'carelessness' misread the Hebrew text and translated it as 'νόμον' (='law'). Copyists made another mistake writing 'ὄνομα' (='name') for 'νόμον'. Thus there are two mistakes, the first leads to the text of the Vulgate, the second to the reading found in some manuscripts of the Septuagint.[2] The proof that the Greek text in its two readings and the Vulgate are corrupt, but that the Hebrew contains the original reading, can, in Reuchlin's view, be found in the works of Jewish grammarians and in the commentaries of Nicolaus of Lyra and Paul of Burgos who depend on the Hebrew wording for their explanations.

Reuchlin finds a confirmation of his proposition that the Seventy translators used a Hebrew text without vowels, in Psalm ci. 24 (cii. 24), where in his interpretation of the word 'koho' he comments on a discrepancy between the Vulgate and Jerome's version. The result of this investigation is found in his *In Septem Psalmos Poenitentiales* when he maintains: 'From this and innumerable other places I conjecture that the Seventy translators had Bibles without vowels.'[3]

---

[1] תּוֹרה and תִּוָּרֵא differ, of course, in their final letters, but תּוֹרא was taken to mean תּוֹרה. Cf. *Biblia Sacra*, ed. R. Kittel–P. Kahle (Stuttgart, 1945; the Psalms are edited by F. Buhl), where the critical note to this word says that Symmachus and Theodotion read תּוֹרא (=תּוֹרָה).

[2] *De Rud. Hebr.* p. 225; *in Sept. Psal. Poen.* fol. k vii^{r–v}.

[3] Fol. k iii^v. Geiger in his book on Reuchlin (p. 136) erroneously writes that the Seventy translators had no Hebrew text, a statement probably due to the omission of one word. O. Kluge, 'Die hebräische Sprachwissenschaft...', *loc. cit.* p. 185 assumes that only Elias Levita drew attention to the history of the writing of vowels in Hebrew.

THE PHILOLOGICAL VIEW: REUCHLIN

One last example may be mentioned here, in which variant readings of a vowel are again shown to underlie a difference between the translations. In Psalm l. 18 (li. 18) the future tense can be understood as a preterite if the vowel belonging to the prefix 'w' is changed. In his *In Septem Psalmos Poenitentiales* Reuchlin refers to *De Rudimentis Hebraicis*, p. 620, explaining that this is not the case in this psalm. Yet the Seventy translators rendered this word into Greek as if it were a past tense, using the aorist with 'ἄν', which according to Priscian's advice had to be translated as a subjunctive mood into Latin. Therefore the Vulgate reads: 'si voluisses...dedissem' (the Authorized Version reads: 'thou desirest no...else would I give'). Reuchlin's version is: 'Quoniam non cupies...et dabo'. This rendering is for Reuchlin a translation according to the 'Hebrew truth'. His method is to examine every word of the Septuagint and of the Vulgate and to compare it with the Hebrew original. In this way he finds inaccuracies and mistakes in the Greek text which cause the Latin text to be wrong. Reuchlin does not even mention that there is a tradition declaring the Seventy translators of the Septuagint to be inspired. He finds out their 'carelessness'. The version made by them cannot therefore serve as a basis of another translation. It is the Hebrew text which alone must be studied. The result of his philological method is the rejection of the translations. But he is also able to praise the Hebrew text. The Jews, he asserts, have preserved the biblical text better than any other nation. In his *Augenspiegel* of 1511 he points out that only the Jews know the number of verses and even of the letters contained in every book of the Old Testament. This hinders them from falsifying the Hebrew wording. Indeed, the different Hebrew manuscripts, whether they are in the orient or in the occident, are identical. Thus their text is correct. This has even been recognized by the Church, for in canon law it is said that the correct text of the Old Testament should be established with the help of Hebrew.[1]

---

[1] *Augenspiegel*, p. xvi^r=Hardt, *op. cit.* p. 30*b*; *Decreta*, pars I, dist. ix, c. vi, quoting St Jerome, *Ep.* LXXI, 5 (*P.L.* vol. XXII, cols. 671–2).

THE PHILOLOGICAL VIEW: REUCHLIN

The authority of the Hebrew text is proved, not only by Reuchlin's philological method but also by the reference to canon law which shows Reuchlin protecting himself against assailants of his claim.[1] For the philological interpretation of the Hebrew Bible it is necessary to learn Hebrew from those who are undoubtedly masters of this language, the Jewish commentators. It is wrong to assume, Reuchlin writes in his *Augenspiegel*, that they purposely falsify the meaning of the Bible. They may be right or wrong in the exegesis of a certain passage, but their interpretation is according to their best knowledge. (In this argument, as so often in his writings, the lawyer Reuchlin introduces a legal interpretation when he determines falsehood as the outcome of bad intentions.) As the Jewish commentators do not misinterpret the text on purpose, they cannot be reproached with falsifying it, although they may make mistakes.[2] It follows that their interpretations cannot be rejected without serious study. Thus the Hebrew text has superiority over all the translations, and the Jewish commentators are preferable to all other expositors since they have a thorough knowledge of the language. It can easily be understood that Reuchlin when rejecting translations was logically bound to recognize the merits of the Jewish commentators.

What is the relationship between Reuchlin's studies and the Schoolmen and the traditional exegesis of the Bible? It can be seen at once that there is no place in Reuchlin's philology for the philosophy of the Schoolmen. He is interested in understanding the linguistic phenomena of God's word in the original Hebrew. This, and only this, is his task. Therefore he need not even mention philosophical questions. They do not exist in his discussions. Philosophy cannot contain the truth since, divorced from God's word as originally given to man, it is only man's garrulity. Reuchlin has expressed this opinion in his first book, *De Verbo Mirifico* of 1494:

You will find many places of Holy Scripture corrupted by ignorant people. In our day the theologians usually attend to Aristotle's dialectic

---
[1] *Augenspiegel*, p. xvi^r–v.   [2] *Augenspiegel*, pp. xvi^v–xvii^r = Hardt, *op. cit.* p. 31 a–b.

sophisms rather than to the words of divine inspiration and of the Holy Spirit. Hence through the study of human inventions even the divine tradition is neglected and man's loquacity extinguishes God's word.[1]

These sentences also illuminate his attitude towards the traditional exegesis. Only the study of the original text can reveal the truth of God's word, for only the text of the Bible in the original language is an unquestionable authority. The truth of anything else needs to be carefully scrutinized. This is expressed in a famous sentence written in 1506: 'Though I revere St Jerome as an angel and though I honour Lyra as a teacher, yet I worship truth as God.'[2] For the discovery of truth Reuchlin did not rely on his own thought or on his personal view. Sebastian Brant, a friend of Reuchlin's, thought this the great mistake of Bible exegesis in his day. Reuchlin is sure that his interpretations are right because, unlike earlier translators who relied on a translation of the Bible, he interpreted the meaning of every word of the original text which was inspired by God. This made him confident in his own power both to understand God's word and to reject medieval commentation, the authors of which relied on the Latin text without knowledge of Greek or Hebrew.

Reuchlin's most important work, as far as the history of Bible interpretation and Bible translation is concerned, is a Hebrew Grammar and Dictionary called *De Rudimentis Hebraicis*. It is in many respects an elementary book, called *Rudimenta* because it is composed 'not for the learned but for those with rudimentary knowledge and those who are to become erudite'.[3] It consists of three parts: the first contains a description of the Hebrew letters and gives advice on how to read, the third is devoted to the study of Grammar whilst the second, the greatest part of the book (pp. 32–541), is filled by a Hebrew-Latin dictionary.

Reuchlin's method in compiling the dictionary is the same as that used in our own time. He gives the equivalent Latin ex-

---

[1] *De Verbo Mirifico* (Basel, 1494) fol. d^r.   [2] *De Rud. Hebr.* p. 549.
[3] P. 3. For all details about *De Rud. Hebr.* see Geiger, *Reuchlin*, pp. 110–33. The dictionary is criticized by O. Kluge, 'Die hebräische Sprachwissenschaft...', *loc. cit.* pp. 93–6.

THE PHILOLOGICAL VIEW: REUCHLIN

pression, generally more than one, for every Hebrew word and then adduces examples for each meaning. These examples are naturally taken from the Old Testament; they are not quoted in Hebrew but in the Latin of the Vulgate. However, Reuchlin often does not agree with this text and in these cases he does not hesitate to say so and to add a new rendering of his own. It is obvious that in his explanations of words he trusts Jewish grammarians and follows them, for he says in the *Augenspiegel*:

For they say how every word in the Bible should be understood according to the special character of their language. In this way Abraham aben Esra, Moses ben Gabirol and Rabbi David Kimhi interpret the words according to grammar....[1]

Through their explanations of words these Jewish grammarians influenced Reuchlin's rendering, and Reuchlin acknowledges his indebtedness to them, especially to David Kimhi who lived from 1150 to 1235. Reuchlin's remarks on the Latin Bible are noteworthy for the outspoken manner in which he criticizes St Jerome's rendering and, though only once, St Augustine's interpretation. These words, typical of Reuchlin's method and of his attitude towards Bible tradition, may be quoted here. When explaining the Hebrew word 'šmt', i.e. 'remisit', 'declinavit', he mentions Psalm cxl. 6 (cxli. 6) quoting the Vulgate: 'Absorpti sunt iuncti petrae iudices eorum' which is rendered in the Rheims-Douay edition: 'their judges falling upon the rock, have been swallowed up'. After the quotation from the Vulgate Reuchlin continues:

I do not know what this translation babbles. The Hebrew truth, however, reads as follows: 'quoniam declinaverunt propter petram iudices eorum'[2] [i.e. 'for their judges have turned aside because of the rock']; some of the Hebrew teachers understand this as relating to Moses and

---

[1] P. xiij^{r-v}=Hardt, *loc. cit.* p. 29b.

[2] Agostino Giustiniano in his polyglot edition of the Psalter in Hebrew, Greek, Arabic, and Aramaic, containing besides the Latin of the Vulgate, two word-for-word translations from Hebrew and Aramaic into Latin (printed P. P. Porrus, Geneua, 1 August 1516) renders the Hebrew text with the same words as Reuchlin but leaves out the word 'quoniam'.

Aaron, the judges of the children of Israel; these turned aside from the faith because of the rock. But the blessed Augustine understands this verse as referring to Aristotle. Moved by some dream he writes that Aristotle trembles in the infernal regions.[1]

Without going into the question about the merits of Reuchlin's criticism it is clear that his words are utterly alien to medieval religious thought. Reuchlin not only praises Jewish learning but he challenges the accuracy of the official Psalm translation and refuses St Augustine's exegesis. He thus openly subjects the authentic version of the Church as well as an explanation of St Augustine's to an examination and rejects both. This is not the only place where Reuchlin criticizes the Vulgate. There is scarcely a page of his dictionary where he does not censure its translation.[2]

This is done by a philological method which traces the meaning of every word in the original Hebrew. No theological argument is ever used by Reuchlin. He is no theologian and openly admits this: 'I do not probe the meaning as a theologian but discuss the words as does the grammarian', he says in his dictionary.[3] This thought epitomises his attitude to philological studies. Though the same view had been expressed before in Erasmus's writings,[4] Reuchlin was the first to apply it to Hebrew studies. Reuchlin mentions this idea again, in a letter written on 10 October 1508:

I would have you know that nobody of the Latin people has been able to give an exact explanation of the Old Testament without having first had knowledge of the language in which it was written.[5]

---

[1] *De Rud. Hebr.* p. 523. The last sentence refers to August. *Enarrationes in Psalmos* (*P.L.* vol. XXXVII, col. 1828). The Vulgate text of today is the same as quoted by Reuchlin. A.V. reads (Psalm CXLI. 6) 'When their judges are overthrown in stony places....'
[2] Geiger (*Reuchlin*, p. 122 n. 3) gives a list of about 200 corrections of the Vulgate. This list is, however, not complete.
[3] P. 123, s.v. 'הוך'.
[4] Preface to Laurentius Valla, *Adnotationes* (Badius, Paris ⟨c. March 7⟩ 1505), printed in *E.E.* vol. I, no. 182 (113–31).
[5] *R.E.* no. 102, p. 105. There is a serious misprint in Geiger's edition. I therefore quote from J. C. Schelhorn, *Amoenitates Historiae Ecclesiasticae et Literariae* (Frankfurt and Leipzig, 1738), vol. II, p. 595: 'Nam hoc tibi affirmo neminem Latinorum Vetus Testamentum potuisse ad unguem exprimere nisi eius linguae, in qua scriptum fuerit, primo peritus sit.' For 'ad unguem' see Erasmus, *Adagia* (*L.B.* vol. II, col. 216 F).

THE PHILOLOGICAL VIEW: REUCHLIN

This principle establishes philology as an autonomous discipline entitled to discuss the meaning of words in the Bible. It proves that theologians may err as to their fundamental meaning and that it may be for philologists to discover the truth hidden in God's word. And it is Truth that Reuchlin 'worshipped as God'.

The effect of this principle on the validity of the Vulgate cannot be overestimated. It means that the traditional text has to be changed at any place where the rendering does not conform to philological considerations. Conclusions drawn from medieval theologians lose their validity if the philologist can prove the inaccuracy of the text on which the medieval commentator relied. The Vulgate thus loses its value as a basis for theological discussion. It may be used in Church service, but according to Reuchlin it cannot be maintained that it is identical with the original. No translation can ever represent God's word in its purity.

Yet it is safe to assume that neither Reuchlin nor any other humanist desirous of reading the Bible in Greek or Hebrew was aware of the danger of his ideas to the Church. The humanists should not be blamed for this. They saw in the study of languages an opening up of a new path to the understanding of God's word, free from the mistakes of those who, though accepted by the Church, misinterpreted texts through their lack of knowledge of the original languages.

The great teacher St Hilary [Reuchlin writes in the *Augenspiegel*] has written commentaries on Holy Scripture, which are praised and accepted by the entire Church. Yet he is often mistaken in the explanation of words because of his ignorance of Hebrew and because he only has a smattering of Greek, as St Jerome writes....[1]

A remedy for this ignorance is, Reuchlin thinks, to follow the advice of the Constitution of Clement V of 1311, in which it is said that Greek and Hebrew should be taught at every university. Therefore Reuchlin concludes his *Augenspiegel* with the suggestion that two teachers of Hebrew should be installed at every University.[2]

---
[1] *Augenspiegel*, p. xiiij^r=Hardt, *loc. cit.* p. 30*a*.
[2] *Augenspiegel*, p. xx^r=Hardt, *loc. cit.* p. 34*b*.

It is hardly possible to follow Jewish grammarians in their explanations of words without being influenced by the contents of their commentaries. Indeed, in the example quoted above, Reuchlin refers in his discussion of Psalm cxl. 6 (cxli. 6) to the meaning ascribed to this verse by the Jewish exegesis. It was natural for Reuchlin, who trusted the scholarship of the Jews, to study their exegesis and their philosophy. He was thus drawn to the Talmud and the Cabbala.

Before 1512 Reuchlin had no knowledge of the Talmud because he could not purchase a copy.[1] Then he succeeded in acquiring one treatise of the Talmud only.[2] It seems that he did not obtain any part of Bomberg's edition in 15 volumes, published at Venice in 1520–23.

He knew, however, the Cabbala, on which he wrote two books: *De Verbo Mirifico* (probably Thomas Anshelm, Pforzheim, 1494), and *De Arte Cabbalistica*, published by Thomas Anshelm, Hagenau, March 1517. What was it that attracted him to a special study of this work? It is very probable that Pico della Mirandola's cabbalistic studies had a permanent influence on him.[3] However strong these influences may have been, they will not be treated here. Nor will the scope of Reuchlin's cabbalistic investigations be discussed, nor the question whether his view on the Cabbala is correct. The emphasis will be laid on the significance of the Cabbala for Reuchlin's study of Hebrew and his quest for truth.

Reuchlin says: 'Cabbala means perception of God's word by hearing'; it is said: 'Mose kibbel', that means: 'Moses heard and perceived the Law from Sinai'.[4] The first 'Cabbala' was given to Adam after his fall. It contained the promise of the Messiah.

---

[1] *Augenspiegel*, pp. iij^v–iv^v=Hardt, *loc. cit.* p. 22 *a–b*.
[2] K. Christ, 'Die Bibliothek Reuchlins in Pforzheim', *Beiheft zum Zentralblatt für Bibliothekswesen*, no. 52 (Leipzig, 1924), pp. 36–7.
[3] Reuchlin mentions Pico, e.g., *De Rud. Hebr.*, p. 620; *Augenspiegel*, pp. xij^v–xiij^r =Hardt, *loc. cit.* p. 29 *a*; *De Arte Cabbalistica*, p. lii E; translations of *S. Athanasius in librum Psalmorum* (Th. Anshelm, Tübingen, 12 Aug. 1515), fol. A iiii^v. Cf. *E.E.* vol. II, no. 324 (11–12), containing the view of John Fisher, bishop of Rochester; L. Geiger, *Reuchlin*, p. 172. [4] *De Arte Cabb.* p. VI E.

THE PHILOLOGICAL VIEW: REUCHLIN

Other revelations were made by God later. All these messages were handed down by mouth from generation to generation. Though several of them were lost, many were written down later and thus preserved. The collection of these messages is contained in the Cabbala.[1] The Cabbalist leads a 'vita contemplativa' and this contemplation is more suited than anything else to lead to the salvation of the soul, to immortality, to the highest beatitude.[2] The Cabbala contains all the wisdom of the world. In it the Bible is explained in such a way that the mind is elevated to God. Its doctrine is identical with that of Pythagoras. Indeed, Pythagoras had learned Cabbala during his long travels. He 'changed' the word 'Cabbala' into the Greek expression 'philosophy'.[3] There are, it is said, fifty gates of understanding of which all but one were given to Moses by God. All these fifty gates are contained in the divine law of the Jews. The student of the Cabbala exerts himself to reach some of them through every possibility of Bible interpretation.[4]

It seems to me that this is the clue to Reuchlin's study of this esoteric doctrine of the Jews, the Cabbala. In it the truth is contained that is given by God. It is for Reuchlin the source of all human wisdom; the Jews were the first philosophers, the source for Pythagoras and Plato; literary art and medicine originated with the Jews.[5] Jewish philosophy, being the oldest and coming directly from God, necessarily contains that truth which Reuchlin 'worshipped as God'. 'There is nothing in our philosophy', he

[1] *De Arte Cabb.* pp. ix s–xiiii T.   [2] *Ibid.* p. ii F; xv C–G.
[3] *Ibid.* p. xxii A; cf. Preface to Reuchlin's translation of *Athanasius, in librum psalmorum,* fol. A iiii^v.
[4] Reuchlin enumerates these methods of Bible interpretation '...vel sensu literali vel allegorico, per dictiones, vel arithmeticas supputationes, vel geometricas literarum figuras sive descriptas seu transmutatas, vel harmoniae consonantias ex formis characterum, coniunctionibus, separationibus, tortuositate, directione, defectu, superabundantia, minoritate, maioritate, coronatione, clausura, apertura, et ordine resultantes' (*De Art. Cabb.* p. lii 1).
[5] Preface to Reuchlin's translation of Joseph Ezobi, *Lanx Argentea* (Thomas Anshelm, Tübingen, March 1512), fol. a^v; Preface to Reuchlin's translation of Hippocrates, *De Praeparatione Hominis* (Thomas Anshelm, Tübingen, 17 February 1512), fol. Aij^v–Aiij^r. Cf. Luther's opinion on this point in his *Tischreden,* vol. I, no. 1040, p. 524. The view of the dependence of Greek philosophy on Moses is already found in Philo of Alexandria.

says, 'that did not belong to the Jews before.'¹ And more generally this is expressed in the words: 'We Latin people drink from the morass, the Greeks from brooks, the Jews from the wells.'² Reuchlin preferred to go back to the wells, and I assume this is the reason why he spent his life in learning and propagating Hebrew rather than Greek. For in order to understand this old wisdom contained in the Jewish sources, one had to learn the language and had to be able to interpret every one of its aspects. During a discussion of a passage of Ezekiel, Reuchlin expressed this thought clearly when saying:

This alone is the field of true contemplation, the single words of which are single mysteries and the single utterances, syllables, the apexes of the letters and the vowels are full of secret meanings, a fact attested not only by Jews but also by Christians.³

These words, I think, refer to the importance of grammar for the understanding of the mysteries of Hebrew. In his last work on Hebrew grammar, called *De Accentibus et Orthographia Linguae Hebraicae* (Thomas Anshelm, Hagenau, February 1518), a definition of grammar is found which is the synthesis of Reuchlin's grammatical and cabbalistic studies:

It is not a slight thing that the knowledge of grammar brings forth. It is certainly the true and authentic meaning of scripture which, as a rule, is disclosed by the innate property of every word, not to mention at the moment the many secrets after which you may hunt, starting not only from the words but almost from the single apexes of the letters, if you set out under the guidance of the spirit. This is especially true of the Hebrew writings where almost nothing has been handed down, that does not, in a shadowy, allegorical way, commend itself to the spiritual understanding.⁴

---

¹ *De Art. Cabb.* p. xxiii.
² Translation of Hippocrates, *De Praeparatione Hominis*, fol. A ij^v.
³ *De Art. Cabb.* p. xxk.
⁴ *De Accentibus*, p. iii^v. The last sentence of this quotation reads: 'id vel maxime verum est de scripturis hebraicis quibus prope nihil est traditum quod non umbraculi vice spiritali commendetur intelligentiae.' Cf. *ibid.* p. lix^v. (The phrase 'umbraculi vice' may be influenced by Origen, e.g. *In Numeros Homiliae xi* (P.G. vol. XII, col. 648 A): 'Et ita omne hoc quod adumbratum est in lege, quoniam lex umbram habet futurorum bonorum. ...')

THE PHILOLOGICAL VIEW: REUCHLIN

Thus grammar not only teaches the meaning of the words but it also leads to the understanding of the secret meaning of the language and thus to God. The only language which fulfils this task is Hebrew, for, in Reuchlin's opinion, it originated before the confusion of the languages and has kept its purity ever since.[1] Hebrew comes from God, as Origen says, it must be holy and the source of all languages, as St Jerome maintains.[2] It is the language which God has spoken.[3]

Therefore we generally call the language of the Jews holy and we say: the sacred letters were written by God's finger and the sacred names were not invented by man but made by God himself.[4]

Thus Reuchlin has found the language which, being holy, must of necessity contain the truth and therefore lead to God. The examples given above are from various dates of Reuchlin's life from 1494 until 1518. We can see him throughout occupied with the study of this language, 'following the genius consuming him' to find out the secrets of all wisdom. The philological and the mystical tendencies are interrelated: philology supports mysticism, and mysticism furthers philology. It is the coincidence of these tendencies that makes him write the following sentences:

The mediator between God and man was, as we read in the Pentateuch, language, yet not any language but only Hebrew; God wished his secrets to be known to mortal man through Hebrew.[5]

Almost two years later (1510) he asserts:

For when reading Hebrew I seem to see God speaking himself, when I think that this is the language in which God and the angels have told their merits to men from on high. And so I tremble in dread and terror, not, however, without some unspeakable joy that follows this my

---

[1] *De Verbo Mirifico*, fol. c 7ᵛ.
[2] *In Septem Psalmos*, fol. a viiiᵛ. The quotations refer to Origen, *In Numeros Homiliae* (P.G. vol. XII, col. 649A); St Jerome, *Commentariorum in Sophoniam Prophetam liber* (P.L. vol. XXV, col. 1384B).
[3] *De Rud. Hebr.* p. 1. Cf. St Jerome, *Praefatio in Paralipomenon* (P.L. vol. XXVIII, col. 1326A).
[4] *De Verbo Mirifico*, fol. c8ᵛ.
[5] Ep. of 11 October 1508 (R.E. no. 102, p. 105).

THE PHILOLOGICAL VIEW: REUCHLIN

wonder or rather awe, which truly I will name knowledge, about which that divine word says: 'The fear of the Lord is the beginning of knowledge'.[1]

'The fear of the Lord is the beginning of knowledge', or as he says in 1512 quoting a sentence from the Talmud: 'With the thoughtless there is no wisdom when there is no fear; and when there is no knowledge there is no understanding.'[2]

This quotation, I think, shows the two sides of Reuchlin's work, his religious reverence which leads him to the study of the Cabbala and his desire for knowledge and truth which causes him to learn Hebrew. The circle of his studies is completely closed. The interdependence of the two aspects of his studies was such that it is impossible to say that one of them was the prime moving force.

The effect of Reuchlin's Cabbalistic studies on his work on the text of the Vulgate was of a very extensive nature. As mentioned above, Reuchlin considers Hebrew a holy language, the source of all languages, free from impurity, and richer than all the other tongues. God's mystery, he maintains, extends to the very letters and their apexes which all have some secret meaning. It is obvious that every translation from Hebrew destroys the holiness, mystery and purity of God's word. It is therefore impossible to render Hebrew into any other language. Yet Reuchlin has made translations from Scripture in all his books dealing with Hebrew studies. There is a further contradiction in these translations. In the preface to Joseph Ezobi's rendering he asserts that word-for-word translations from Greek to Latin are not permitted, adding that they cannot be made from Hebrew into Latin.[3] All Reuchlin's renderings, however, are word for word. The reason for this

---

[1] *Ep.* of 19 March 1510 (*R.E.* no. 115, p. 123). Reuchlin changes the order of words in Proverbs, i. 7, and reads: 'initium sapientiae timor dei' instead of 'Timor domini principium sapientiae'.

[2] Preface to his translation of Joseph Ezobi, fol. a iij^v where the Hebrew words are explained in the following way: 'significans apud temerarios non esse sapientiam, ubi nihil timetur, et ubi non est scientia ibi nullam affore prudentiam'.

[3] Fol. a iij^r.

cannot be his inability and unwillingness to write in a style that is in accordance with humanistic requirements.[1] The real reason for using the word-for-word translation is mentioned by Reuchlin: he does not translate Scripture for literary purposes. His renderings are intended to be a help to the student. They should guide him to the understanding of the Hebrew text: for this a word-for-word rendering is best suited.[2]

Since according to Reuchlin every translation of the Bible diminishes its value, it is only through the Hebrew text that his aim can be fulfilled, namely to reconstitute the authority of the Old Testament.[3] There is the danger, he says, that the Bible will be completely forgotten. It is held in contempt because of the sophisticated interpretations of the Schoolmen and because of the stylistic tendencies of his time where the rhetoric and loveliness of the poets is esteemed. Both these reasons for the neglect of Scripture can be remedied by leaving the Latin Vulgate and returning to the original Hebrew, for the exegesis of the Latin text is not valid for the Hebrew wording and the stylistic objections to the Vulgate are of no account as the Hebrew Bible is elegant in its diction.[4] Thus the result of Reuchlin's studies is: the Vulgate must perish that the Bible may live.

One example may suffice to show the method of Reuchlin's translations. In this rendering of Psalms cix–cxiv, preserved in a Paris manuscript, he gives five Latin variants for one phrase. The Vulgate of Psalm cix. 2 (cx. 2) reads 'Virgam virtutis tuae...'. Reuchlin translates: (1) 'Virgam potestatis tuae', (2) 'Stipitem roboris tui', (3) 'Baculum firmitatis tuae', (4) 'Sustentatorem virtutis tuae', (5) 'Fulcimentum fortidudinis tuae'. There is an additional note:

---

[1] Reuchlin speaks about his style in *De Rud. Hebr.* p. 546. Cf. his letters to Erasmus of 5 June 1516 and 27 March 1517 (*R.E.* no. 219, pp. 249–50, and no. 237, p. 267, reprinted *E.E.* vol. II, no. 418 (14–21), and no. 562 (1–14)).

[2] *In Septem Psalmos Poenitentiales*. The full title is:...*In Septem Psalmos Poenitentiales hebraicos interpretatio de verbo ad verbum et super eisdem commentarioli sui 'ad discendum linguam hebraicam' ex rudimentis*. Cf. ibid. fol. a vii^v; *Augenspiegel*, p. xxvi^r=Hardt, loc. cit. p. 49b.

[3] *Augenspiegel*, p. xxvi^r=Hardt, loc. cit. p. 49b. *De Accentibus*, iii^r.

[4] *De Rud Hebr.* p. 1; *De Accentibus*, lx^r.

'tui vel tuae in communi translatione graeca non habet sed in hebraica et in 70.' These different renderings prove that none of them is really satisfactory. They seem to be an attempt at interpretation rather than at translation. However this may be, these variants show that Reuchlin does not wish to give one fixed text as a translation for the Hebrew text. This is in complete accordance with his aim of guiding the student to the original text.[1]

Hebrew studies thus result in conclusions dangerous to the medieval church:

(1) The text of the Vulgate and the commentaries based on it are questioned and re-examined. The use of Hebrew grammarians and expositors for this investigation easily evokes suspicions about the Christian interpretations.

(2) Every translation of Scripture impairs or even destroys its real and secret meaning. Thus the authority of the Vulgate, the authoritative version of the Church, is directly challenged.

Yet Reuchlin was opposed to every aspect of Luther's doctrine. He died a member of the Roman Catholic Church. His aim was, to use a phrase of Erasmus, 'to restore the old',[2] and he did not see the danger that this implied for the Church of his time.

The impact of Reuchlin's teaching was strongly resented by those who followed the 'traditional view'. The humanists, as adherents of a new movement that had to fight for its existence, always took strong exception to any attack made upon them and used to answer their opponents with great indignation and acrimony. This is not the place to discuss the details of the Reuchlin controversy.[3] Before pointing out the effect of this quarrel upon the

---

[1] The MS. is in Paris (Latin 7455, pp. 274–7). The passage quoted is on p. 274a. It is not written by Reuchlin. As the translation is not fixed and without any final text it is very unlikely that it was intended for publication. Geiger, *Reuchlin* (p. 139), may well be right in his assumption that it is a record of one of Reuchlin's lectures.

[2] *E.E.* vol. IV, no. 1153 (185–6), of 18 Oct. 1520.

[3] All the dates concerning the controversy are given by E. Boecking, *Epistolae Obscurorum Virorum* (*Ulrichi Hutteni Equitis Operum Supplementum*), vol. II, pp. 117–56. A description is found in every historical work of the time. Geiger (*Reuchlin*, pp. 205–454) gives a detailed though very biased description. In English little has been written about

authority of the Vulgate it may be worth while to mention that some of the humanists criticized his cabbalistic studies. Erasmus states several times in his letters that the Cabbala does not please him.[1] John Colet, Dean of St Paul's, mistrusted these studies of Reuchlin, admitting his ignorance and 'blindness in matters so remote and in the works of so great a man', but expressing his opinion that the goal of life was to strive after purification, illumination and perfection. Reuchlin, Colet writes, promises to achieve this through Pythagoras and the Cabbala; he himself, however, believes in one way only: 'ardent love and imitation of Christ'.[2] Colet apparently sees the danger of Reuchlin's cabbalistic studies in the fact that they lead to a belief in salvation outside Christianity. This is a serious point against Reuchlin. It is used by the inquisitor Jacob Hochstraten in whose book *Destructio Cabbalae* the reproach resounds: 'Reuchlin is a herald of Cabbalistic perfidy': an attack rejected by Erasmus in a letter to Hochstraten.[3] The inquisitor's book was only one example of the many charges against Reuchlin.

What are these charges? It must be remembered that there was no answer to the principle that every interpretation must be derived from the original text. Reuchlin's opponents had to face the very awkward question whether Hebrew studies should be forbidden. But they could not attack him on this account, for there was the advice of Pope Clement of 1311, that Hebrew should be studied. Besides, some of Reuchlin's opponents who had sympathies with the humanists wished, I believe, to avoid a conflict on this issue.

it. The Preface and Introduction of F. G. Stokes, *Epistolae Obscurorum Virorum, The Latin text with an English rendering*...(London, 1909), pp. vii–lxxiii gives the best account. It is reviewed by Sir Adolphus Ward, *loc. cit.* pp. 86–117.

[1] For Erasmus's attitude see *E.E.* vol. III, no. 967 (71), of 18 May ⟨1519⟩; vol. IV, no. 1033 (35–6), of 19 October 1519; no. 1155 (18), of 8 November 1520; no. 1160 (1–5), of ⟨November⟩ 1520. After Reuchlin's death Erasmus praised him in his apotheosis 'De incomparabili heroe Johanne Reuchlino in divinorum numerum relato' in *Colloquia Familiaria* (L.B. vol. I, cols. 689–92).

[2] Colet's letter is printed *E.E.* vol. II, no. 593. It is mentioned by Geiger, *Reuchlin*, p. 198. Cf. Erasmus's letter to Reuchlin of 29 September ⟨1516⟩: 'Ioanni Coleto sacrum est tuum nomen' (*E.E.* vol. II, no. 471 (15)).

[3] *Destructio Cabalae seu Cabalisticae perfidiae* (Quentell, Cologne, April 1519). Erasmus's letter is *E.E.* vol. IV, no. 1006; see esp. lines 89ff.

THE PHILOLOGICAL VIEW: REUCHLIN

Before the controversy started Reuchlin, well aware of the danger of his Hebrew studies for himself, had written in his *De Rudimentis Hebraicis* in 1506:

Enemies will rise against my dictionary in which the translations of many are frequently censured. 'Oh! what a crime', they will exclaim. 'Nothing is more unworthy of the memory of the Fathers, no crime more cruel, since that most audacious man strives to overthrow so many and such saintly men who were inspired by the divine spirit. As Pope Gelasius attests, the Bible of the most blessed Jerome was accepted in the Church. The venerable father Nicolaus of Lyra, the common expositor of the Bible, is approved as an irreproachable man by all faithful to Christ. Now a certain puff of smoke, a little Reuchlin, has appeared,[1] who indicates that these have ignorantly translated in a great many places.' I answer these threatening shouts with these few words: to allow me what was allowed to these very famous luminaries.[2]

The permission for which Reuchlin asks is mentioned in the following lines: St Jerome criticized earlier translators and Nicolaus of Lyra censured the renderings of St Jerome. But 'St Jerome himself admits his own mistakes in his renderings'.[3]

Reuchlin's fear seemed unfounded. Nothing happened for some time. When, however, he was asked by the Emperor Maximilian to advise whether the Talmud should be burned, the storm broke. From an anti-semitic attack made by the proselyte Pfefferkorn, a controversy started in which the fundamentals of humanism were attacked. Only this aspect of the quarrel shall be outlined here.

One of the teachers of theology at the University of Cologne, Collin, wrote on 2 January 1512, to Reuchlin: 'It is not surprising

---

[1] This is a pun on Reuchlin's name which Reuchlin himself thought to be a diminutive of 'Rauch', i.e. 'smoke'. It is used here in a derogative meaning and translated as 'fumulus'. The text reads: 'iamiam exortus est aliquis fumulus....' He generally calls himself 'Reuchlin' but also in a Greek translation of 'fumulus' '*kapnion*'.

[2] *De Rud. Hebr.* p. 548. There is some similarity of thought in these sentences with Erasmus's Preface to Laurentius Valla, *Adnotationes* of 1505, reprinted *E.E.* vol. I, no. 182 (107ff.).

[3] *De Rud. Hebr.* p. 548 refers to St Jerome, *Commentaria in Isaiam*, xix. 16–17 (P.L. vol. XXIV, col. 184).

that a lawyer does not understand the subtleties of theology.'[1] This argument is repeated in Arnold of Tungern's book against Reuchlin, *Articuli sive Propositiones de Iudaico Favore nimis Suspectae*, of 28 August 1512.[2] And again, Hochstraten repeats this charge: Reuchlin is a 'hater of theology and philosophy',[3] 'a herald of cabbalistic perfidy' who endeavours to extol the fabrication of the Jews in order to execrate the studies of theology and to eliminate the practice of syllogisms.[4] Mockingly, Hochstraten calls the followers of Reuchlin 'not only philosophers and lawyers but also theologians, the unconquered defenders of the faith'.[5]

This is a very serious attack delivered exactly where it had been anticipated in Reuchlin's *De Rudimentis Hebraicis* of 1506. It means that nobody who is not a theologian is allowed to criticize matters concerning theology. Reuchlin therefore as a philologist is not entitled to censure the interpretations and translations of theologians since 'he cannot understand the subtleties of theology'. Thus the linguistic investigations of Reuchlin are of no value. His results must of necessity be wrong. This is the denial of Reuchlin's demand for the same freedom of criticism that was allowed to Nicolaus of Lyra and to St Jerome. It is the rejection of Reuchlin's claim for intellectual freedom, the freedom to search for truth unhampered by the fetters of tradition and authority.

Reuchlin understood the menace to humanism at once when in 1512 he received the above-mentioned letter from Collin. His first answer was given in March 1512, when he still hoped to stave off the attack and to keep the peace. He mentioned that his fellow humanists would come to his defence.[6] In later years he always drew attention to the fact that the attack was not against him personally but against humanism. He spoke of the University of Cologne as 'a species of most inhuman men who call themselves

---

[1] R.E. no. 134, p. 150: 'Non mirum si iurista theologicas non attigerit subtilitates.'
[2] Published by Quentell, Cologne; see esp. fol. B ij^r.
[3] *Apologia* (Quentell, Cologne, 1518), fol. a iiij^r.
[4] *Destructio Cabalae*, fol. i^v (leaf 1).
[5] *Apologia Secunda* (Quentell, Cologne, 1 October 1519), fol. F iij^v.
[6] R.E. no. 144, pp. 166–7 (dated 11 March 1512).

theologians'. These had written against several lawyers, and against all humanists, before they had turned against him. His activities as a propagator of Hebrew were the real reason for their hatred towards him.[1] And more than a year later, in November 1514, he wrote in the same vein, namely that if his adversaries succeeded in destroying him, they would try to subdue Erasmus and one by one all the humanists.[2]

The humanists composed the *Epistolae Obscurorum Virorum* which made their opponents ridiculous.[3] When the Pope decided against Reuchlin in 1520, this condemnation of Reuchlin's book *Der Augenspiegel* did not remove the ridicule and the ignominy to which the opponents of humanism were exposed. The men who fought Reuchlin have kept their name of 'obscurants' ever since.

This means that their attack on Reuchlin had utterly failed and that Hebrew studies could be developed without great interference from the side of the 'obscurants'.[4] Indeed, it means that Reuchlin was condemned only because his work was thought to be favourable to Luther, for there was no objection against the Talmud in Rome, where the pope gave privileges to the printer Bomberg at Venice to print the complete Talmud for the first time.

The universities that had been in the forefront of the attack against Reuchlin suffered a moral blow, for the repetition of their condemnatory judgements inevitably lowered their prestige. Their antipathy to any new tendency was well known and everybody who did not accept tradition could expect to get into difficulties with them. This, of course, blunted the power of their decisions

---

[1] R.E. no. 171, pp. 198–9; for text see *Illustrium Virorum Epistolae*, fol. v ii^r.

[2] R.E. no. 198, p. 231; for text see G. Friedlaender, *Beiträge zur Reformationsgeschichte. Sammlung ungedruckter Briefe des Reuchlin, Beza und Bullinger* (Berlin, 1837), p. 47: 'Lovanii fertur, hoc adversariis esse constitutum, ut si me oppresserint, Erasmum Roterodamum sint aggressuri, et ita singillatim omnes se velle poetas (sic enim bonarum literarum studiosos appellant) eradicare.'

[3] The support of the humanists is described by Geiger, *Reuchlin*, pp. 321 ff. An interesting list of friends of Reuchlin in England is printed R.E. no. 226, p. 259 n. 5 (*E.E.* vol. II, no. 471).

[4] In 1542 Johannes Eck still writes against the Talmud and Hebrew studies in *Ains Juden büechlins verlegung*... (Ingoldstadt, 1542), fols. O iij^v–Qiij^v, esp. fol. P iv^v–Qi^r.

and made it easier to ridicule them as obscurants. Reuchlin in the letter mentioned above[1] speaks of attacks against lawyers and humanists. Luther was more explicit after his condemnation by the Universities of Louvain and Cologne. In his *Responsio ad Condemnationem doctrinalem per Magistros Nostros Lovanienses et Colonienses Factam* he enumerated those who had been wrongly condemned by universities, among them Occam, Johannes Pico della Mirandola and Laurentius Valla. He continued that he wished to omit others who had been attacked, namely Faber Stapulensis and Erasmus. Luther asserted that the universities, conscious of their ignominy as a result of their attack on Reuchlin, assailed him that they might regain their reputation.[2] Luther interpreted the acts of the universities as being concerned with the furtherance of their own interests and not with the establishment of truth. That he could do so, proves that the position of the universities was morally undermined. Their opposition to Luther was further weakened through their fight against Erasmus, who in pursuit of Greek studies had published a new translation of the New Testament in 1516.

[1] R.E. no. 171, pp. 198–9; for text see *Illustrium Virorum Epistolae*, fol. v ii^r.
[2] W.A. vol. VI, pp. 183–4.

CHAPTER V

THE PHILOLOGICAL VIEW: ERASMUS OF ROTTERDAM

While the study of Hebrew necessarily leads to theological research, especially to the Old Testament in its original language, Greek studies may easily be limited to literature, history, philosophy, and other non-theological subjects. An indication of the tendency to use the knowledge of Greek for profane subjects can be seen in the earliest printed books, for among sixty-three Greek books published before 1500 there is no Greek New Testament and only three Psalters, one of 20 September 1481 (Mediolani), the second of 15 November 1486 (Venice), and the third not later than the middle of 1498 (Aldus, Venice).[1]

The lack of interest in Greek studies in Europe outside Italy can be learned from the fact that only in Italy were Greek books printed until 1507, when Theocritus came out in Paris. The first Greek publications in other countries north of the Alps were even later.[2] 'With Aldus Manutius a fresh period in the history of printing opens.' Aldus's first edition was K. Lascaris's *Erotemata* of February/March 1495 the grammar of which was often reprinted in Germany from 1501 onwards.[3]

Before 1500 only very few humanists north of the Alps could boast a knowledge of Greek. There was hardly ever an opportunity

[1] R. Proctor, *The Printing of Greek in the Fifteenth Century* (Oxford, 1900), pp. 49–52.
[2] It may be useful to list the first editions of books wholly in Greek in European countries: FRANCE: 1507, printer Gilles de Gourmont, Paris, containing Theocritus. GERMANY: probably 1513, printer Grunenberg, Wittenberg, containing *Batrachomyomachia*. SPAIN: 1514, printer Arnold of Brocario, Alcala, a polyglot New Testament (Complutensian). ENGLAND: probably 1543, printer Reg. Wolfe, London, containing Chrysostom, *Two Homilies*. See Proctor, *op. cit.* pp. 140ff.; F. Madan, *Oxford Books*, vol. I (Oxford, 1895), p. 18, no. 5 n.; H. R. Plomer, *A Short History of English Printing 1476–1900* (London, 1915), p. 85. The first book containing some Greek characters is Cicero, *De Officiis, Paradoxa*, printer Fust and Schoefer, Mainz 1465, where the headings are printed in a very very strange mixture of Greek and Latin types. For details see R. Proctor, *op. cit.* pp. 24–6.
[3] R. Proctor, *op. cit.* pp. 24, 93; cf. A. Horawitz, *Griechische Studien, Beitraege zur Geschichte des Griechischen in Deutschland* (Berlin, 1884), vol. I, pp. 8ff.

THE PHILOLOGICAL VIEW: ERASMUS OF ROTTERDAM

of learning it. Grammars were elementary and the teaching, even if done by Greeks, was of doubtful quality. There was Hermonymus of Sparta, who stayed in England for a short period in 1475/6 before he went to live in Paris. He was, it seems, the only teacher of Greek for a considerable time. Among his pupils were Reuchlin, Erasmus, Budaeus, Beatus Rhenanus and Michael Hummelberg, who were critical of his knowledge of Greek grammar.[1] Erasmus of Rotterdam, for example, had considerable difficulty in finding a suitable teacher. At Deventer, which he left in 1484, he had learnt the bare rudiments from Hegius, whose pupil he was for a short time. He had therefore to take up the study of Greek in later life. At the beginning of 1501 he had a teacher, probably Hermonymus, 'in 1502 he had become fluent in the tongue', and only in 1504 could he freely use it for his studies.[2]

Though the early humanists of the fifteenth century would have liked to learn Greek, their whole strength and energy was absorbed in evolving a Latin style which would be elegant and in accordance with the rules of Latin grammar.[3] Another reason which militated against the study of Greek became of importance mainly at the beginning of the sixteenth century. It was the fear that a language used by the heretical Greeks must contain heretical views. Indeed, the humanists were accused of heresy because of their knowledge of Greek.[4] This reproach was most loudly voiced after the humanists had turned their attention to the Greek Bible.

---

[1] On Hermonymus as a teacher see *E.E.* vol. I, p. 7, l. 22 note. Greek knowledge in England is discussed by R. Weiss, *Humanism in England during the Fifteenth Century* (Oxford, 1941), pp. 142 ff.; cf. P. S. Allen, *The Age of Erasmus* (Oxford, 1914), pp. 120 ff. For Germany see O. Kluge, 'Die Griechischen Studien in Renaissance und Humanismus', *Zeitschrift für Geschichte der Erziehung und des Unterrichts*, vol. XXIV (1934). The history of Greek studies in Basel is described by R. Wackernagel, *Geschichte der Stadt Basel* (Basel, 1907–24), vol. II, pp. 592–3; vol. III, pp. 215–17; cf. *E.E.* vol. I, p. 4, ll. 29–31.

[2] *E.E.* vol. I, Appendix VI, pp. 592–3. To the references given there add *E.E.* vol. I, no. 181 (29–34).

[3] For a short survey of German humanism see W. Schwarz, 'Translation into German in the Fifteenth Century', *Modern Language Review* (1944), pp. 368 ff.

[4] For some examples see O. Kluge, *loc. cit.* pp. 12–13. Reuchlin, *de Accentibus*, 1512, p. iii^r. Cf. *E.E.* vol. III, no. 843 (342–9), of 7 May 1518; *E.E.* vol. V, no. 1334 (834), of 5 January 1522/3.

THE PHILOLOGICAL VIEW: ERASMUS OF ROTTERDAM

It should, however, not be assumed that the Church was officially opposed to Greek studies or to editions of the Greek Bible. On the contrary, Leo X, who was pope from 1513 to 1521, was most favourably disposed towards these endeavours. The first polyglot Bibles were dedicated to him, and he even lent Greek manuscripts from the Vatican to one of the editors.[1]

In this chapter the work of one humanist only can be described, but he was undoubtedly the greatest at the beginning of the sixteenth century—Erasmus of Rotterdam. His 1516 edition of the New Testament is an important event in the history of Biblical studies since it is the first complete edition ever to be published with a Greek text and a translation based on it. His life bears witness to the fact that the use of Greek for the study of Holy Writ was considered dangerous. Indeed he was the first who, against strong opposition, established the bond between the two. It is for this reason that the development of his thought must be described in a history of the study of the Bible. Instead of discussing his 1516 edition of the Bible an attempt will be made to show that this edition was the outcome of a slow progress in which he blended contemporary humanism and theology according to his lights.

The life of Erasmus can be followed in great detail. All the outer circumstances of at least his first thirty years of his life drove him towards monastic life and the study of theology. In 1520 looking back at this time of his life he said 'no judgement (for I was too young to judge) but natural inclination rapt me away as if inspired to the temple of the Muses'.[2] And again in 1523: 'velut occulta naturae vi rapiebar ad bonas literas'. These two passages remind us of Reuchlin's: 'ingenio trahor et amore pietatis.'

---

[1] Polyglots dedicated to Pope Leo X are: *Psalterium octaplum*, ed. Agustino Giustiniano, of 1516; the *Complutensian Bible*, printed between 1514 and 1517 (for this edition Leo X had sent a MS. to Spain; see P. S. Allen, *Erasmus Lectures and Wayfaring Sketches* (Oxford, 1934), pp. 142–3); Erasmus's *New Testament* of 1516.

[2] *E.E.* vol. IV, no. 1110 (6–7).

THE PHILOLOGICAL VIEW: ERASMUS OF ROTTERDAM

It is necessary to pause here and to see the serious limitations imposed on our knowledge of Erasmus's thought. Some letters and some early works have been preserved. Yet Erasmus's sayings are often contradictory. It is possible to find remarks on the same subject written at a short interval which seem to indicate a complete reversal of his thought. The elegance of his style and the polish of his language frequently conceal rather than reveal his views. Moreover, the number of letters relating to his earlier life is restricted. In some cases, as for example in the exchange of letters with Colet, letters were a continuation of conversations and it is impossible to understand the allusions and special significance of certain words. These factors should make the student cautious. Our knowledge does not allow us to trace Erasmus's development with certainty. It would, however, be rash to assume that his letters are insincere because of inconsistencies. It is more likely that Erasmus's view was not yet rigidly fixed and that some of his inconsistencies are due to his own vacillations. In this study I shall stress those points in his early writings which can, in a maturer form, be found in his later works, and in this way attempt a rough outline of his development. Obviously, the date of the inception of a new thought is not determined by its first occurrence in writing, which provides only a *terminus ad quem*. But we who live about 450 years after these letters were composed must be content to ascribe it to the time when documentary evidence is available. If we do not follow this principle, we give free rein to our imagination.

With these limitations in mind we may say that Erasmus was 'rapt away' to *bonae literae* which to him meant the study of literary works.[1] The first aim was to gain mastery of Latin. Greek did not come yet into his orbit. In some of his early letters, written in 1489, he advises his friends which authors they should read. These

[1] The meaning of *bonae literae* for Erasmus has been illuminated by R. Pfeiffer, 'Humanitas Erasmiana', *Studien der Bibliothek Warburg*, vol. XXII (1931), esp. p. 7. A list of Erasmus's variations of this expression is given by P. S. Allen, 'The Humanities', *Proceedings of the Classical Association*, vol. XIV (1917), p. 130, where (pp. 128–9) the impossibility of translating terms like 'humanitas' into English is stressed.

lists contain the names of poets generally known to the fifteenth-century humanists: Vergil, Horace, Ovid, Juvenal, Statius, Martial, Claudian, Persius, Lucan, Tibullus, Propertius, Sidonius, and writers such as Cicero, Quintilian, Sallust, Terence (who at that time was considered a prose writer). Once when enumerating authors of excellent precepts on rhetoric he mentions Geoffrey de Vino Salvo (Vinsauf) who lived at the beginning of the twelfth century. He also records lists of Italian and German humanists.[1] All these authors are his 'guides' ('duces') but the crown is given to Laurentius Valla whom he, quoting Ennius from Cicero or Quintilian, calls 'suadae medulla' and 'Attica musa',[2] 'the marrow of persuasion' and 'Attic muse'. Valla, of whose *Elegantiae* Erasmus made a paraphrase,[3] is for Erasmus the writer who more than any other can be relied upon to observe the elegancies of language.[4] Erasmus refers to Valla's theory that the decay of language occurred at the same time as the decline of the other arts. In the olden times, Erasmus points out, arts and eloquence were flourishing, but through the obstinacy of the barbarians this civilization disappeared without a trace. Teachers who themselves had learned nothing taught their pupils nothing but the latest precepts.[5] Their pupils having learned all this 'nonsense' were unable to say even one Latin sentence. Laurentius Valla and Philelphus hindered these 'barbarians' from destroying the Muses completely.[6]

These thoughts, although dependent in essence on Valla, contain three points important for the understanding of Erasmus. First, in Erasmus's view, a new epoch had started which was different from the immediately preceding period which we call

---

[1] All these letters were written in 1489: *E.E.* vol. I, nos. 20 (96–105), 27 (42–50), 31 (83–5).
[2] *E.E.* vol. I, no. 29 (7 and 42).   [3] *E.E.* vol. I, no. 23 (106 n.).
[4] *E.E.* vol. I, no. 20 (101–2).
[5] Cf. *Antibarbari*, ed. A. Hyma, *The Youth of Erasmus* (Ann Arbor, 1930), pp. 287 ff. = L.B. vol. x, cols. 1715 ff. For the date of *Antibarbari* see *E.E.* vol. I, no. 30 (16 n.). R. Pfeiffer, 'Die Wandlungen der Antibarbari', *Gedenkschrift zum 400. Todestage des Erasmus von Rotterdam* (Basel, 1936), pp. 62, 65 ff.
[6] *E.E.* vol. I, no. 23 (78–106).

the Middle Ages. He dates the beginning of this new age of humanism from Valla and Philelphus. At this time of his life he seems to be ignorant of Petrarch.[1] Secondly, he has in mind three classes of barbarians: those who brought about the destruction of the civilization of Antiquity without leaving a trace of it; those who, themselves illiterate, teach what they do not know, making their pupils ignorant; and those who have turned away from the precepts of old and invented new rules which are sheer madness. Valla 'exposed the absurdities of the barbarians and brought back into use the observances of orators and poets long covered with the dust of oblivion'.[2] Therefore an attack on Valla is for Erasmus an attack on humanism: 'Only barbarians dislike him.'[3] Thirdly, human beings, namely the barbarians, destroyed the old civilization. Its revival can be achieved by human endeavour. This is a most significant view, which Erasmus retained all through his life. It is within the capacity of man to change things. He has the free will to do so. He is not dependent on an outside force that guides and rules him. Erasmus points this out in a book called *Antibarbari*, part of which was probably written at the same period as the letters quoted above. He maintains that the decay of civilization is due neither to the influence of the stars, nor to the coming of Christianity although some believe that 'religion and civilization do not go well together', nor to the old age of the earth causing a general degeneration of man. It is man's fault. It is due to the

---

[1] This is not mentioned by H. Weisinger, 'Who began the Revival of Learning? The Renaissance Point of View', *Papers of the Michigan Academy of Science, Arts and Letters*, vol. xxx (1944) 1945, pp. 625 ff.

[2] E.E. vol. I, no. 23 (81–106) = N. vol. I, p. 67; E.E. vol. I, no. 26 (63–7, 103–8).

[3] E.E. vol. I, no. 29 (49 and 32–4) = N. vol. I, p. 73. E.E. vol. I, no. 30 (20 ff.) distinguishes between two groups of barbarians, cf. *Antibarbari*, ed. A. Hyma, *The Youth of Erasmus*, pp. 271–3 = L.B. vol. x, cols. 1704 ff. The connexion between the letters and *Antibarbari* where he also characterizes the barbarians is discussed by R. Pfeiffer, 'Die Wandlungen der Antibarbari', *loc. cit.* pp. 65 ff. There is a different definition of 'barbari' in E.E. vol. I, no. 26 (25 ff.) which is of no importance in this connexion. Erasmus uses the word for the first time in E.E. vol. I, no. 22 (23). Medieval commentators and grammarians like Papia, Huguitio, Ebrardus listed by Erasmus in his letters are among the barbarians, see E.E. vol. I, nos. 26 (88–90), 31 (48–51, 80–4); cf. *Antibarbari*, loc. cit. pp. 266 and 289 = L.B. vol. x, cols. 1701, 1716. Cf. Jacob Wimpfeling, *Isidoneus Germanicus* (J. Grüninger, Strassburg, c. 1500), pp. xvi^r (beginning of ch. 21) and xxiiii^r (end of ch. 26).

stupidity of teachers who corrupt the intelligence of their pupils. These teachers are, of course, barbarians, and the fight to replace them by erudite scholars is the first step to a new civilization.[1]

In this fight the humanists used, of necessity, arguments taken not only from profane literature but also from authorities of the Church, whose validity even the 'barbarians' could not deny. Jerome was quoted by Erasmus in his early letters as a witness against the barbarians. 'If they [the barbarians] fairly looked at the Epistles of Jerome, they would understand that dulness is not sanctity, nor elegance of language impiety.'[2] It is characteristic of Erasmus that he used St Jerome's letters only for the defence of humanism. He himself had copied them and probably knew some of Jerome's works. He was interested in Augustine's writings,[3] and he may have been conversant with many other Fathers of the Church. But it is important to realize that at this period of his life his interest in them as expressed in his writings was to prove that they favoured the humanities and to find 'darts with which to refute the reproaches of the barbarians'.[4] For this reason he passed under review some parts of Jerome's and Augustine's work in the *Antibarbari*. He was able to show that these Fathers recommend pagan literature to Christians for reading.[5] They had pointed out that the Bible supported their view in this point: Moses and Daniel knew the science of Egypt and Chaldea. The spoils taken on God's command by the Jews when they left Egypt signify that the knowledge of the pagans should be used for the exegesis of the Bible.[6] There is therefore no reason for neglecting the writings of secular pagan authors. There are people, Erasmus continues, who assert that learning makes man proud. Yet this is not true, for those who are really learned are modest.[7] Many objections to the humanities may be taken from a gloss. To this attack Erasmus

---

[1] *Antibarbari*, ed. Hyma, *loc. cit.* pp. 249 ff. = *L.B.* vol. x, cols. 1695 c f.
[2] *E.E.* vol. I, no. 22 (18–19 and 21–2) = *N.* vol. I, p. 75, of ⟨June 1489?⟩.
[3] See P. S. Allen in *E.E.* Appendix V, vol. I, p. 590.
[4] *E.E.* vol. I, no. 22 (22–4).
[5] *Antibarbari*, ed. Hyma, *loc. cit.* pp. 309 f. = *L.B.* vol. x, col. 1729 F.
[6] *Antibarbari*, ed. Hyma, p. 313 = *L.B.* vol. x, cols. 1732 c ff.
[7] *Antibarbari*, ed. Hyma, pp. 285–7 = *L.B.* vol. x, cols. 1713–15.

THE PHILOLOGICAL VIEW: ERASMUS OF ROTTERDAM

replies sarcastically that these glosses are taken from 'barbarous authorities'.[1] Thus the last blow is struck: the glosses, that means the authorities of the 'barbarians', are worthless because they lack erudition. The barbarian can prove his view only with the help of the authority of another barbarian. It is necessary to return to the early Fathers of the Church and thus to get rid of every kind of barbarism.[2] Erasmus's defence of humanism goes together with his attack upon the 'barbarians'.

The basis from which a maturer and more developed thought will be derived is clearly recognizable in these early writings. Erasmus trusts in the capacity of man to create a new civilization by his own exertions. When analysing the past, man is not only capable of seeing the reasons for the decline of civilization but he is also equal to the task of bringing to new life the great literary works of the past and of following the precepts advocated in them. The civilization thus created will be new since it disregards the achievements of the immediate past, and it will be old since it revives and reinstates ancient thought. This aim could be accomplished only through erudition, not through hope of, nor reliance on, inspiration. Erasmus does not deny the existence of inspiration. But, quoting Augustine's *De Doctrina Christiana*, he denies that inspiration can replace man's work.[3]

Although he was aware of the importance of Greek civilization[4] Erasmus concentrated on learning the precepts of Latin antiquity for use in his own writings. Indeed, his later statement is true: he was 'rapt away' to *bonae literae*. In these years he acquired a mastery of Latin unparalleled among contemporary humanists north of the Alps. Thus he could proceed to a thorough training in Greek. He well knew that 'almost everything worth knowing was written in these two languages'. Therefore in 1497 in his *Ratio*

---

[1] *Antibarbari*, ed. Hyma, pp. 310, 324–5 = L.B. vol. x, cols. 1730c, 1739ff.
[2] *Antibarbari*, ed. Hyma, pp. 291–2 = L.B. vol. x, cols. 1717F–1718D.
[3] *Antibarbari*, ed. Hyma, p. 325 = L.B. vol. x, cols. 1740D–E. Cf. *P.L.* vol. xxxiv, cols. 15ff., esp. col. 16.
[4] See, for example, *Antibarbari*, ed. Hyma, p. 283 = L.B. vol. x, col. 1712F: 'Artes studiosa Graecia repperit....'

THE PHILOLOGICAL VIEW: ERASMUS OF ROTTERDAM

Studii he speaks about the importance of learning Greek and Latin. In this book, in which no reference is made to Christian authors, the student is warned against the Schoolmen and advised of the great importance of a good style.[1]

The study of Greek was to give a new impetus to Erasmus's bent towards *bonae literae*, and in all probability he learned this language for the sake of the literature. There is no remark in Erasmus's writings of this period in which he expresses the belief that Greek language would be of importance for the understanding of Holy Writ.

Theology, it seems, had lost every attraction for Erasmus. If it were so, it could easily be explained. He lived and studied in Paris, the centre of scholasticism. Here he learned, or should have learned, a philosophy which he despised. It was in these years at Paris that he wrote:

The student should not learn logic from that most loquacious kind of sophists, he should not sit at their feet for long and should not get old there, at the rocks of the Sirens, as it were.[2]

These Schoolmen who, as he wrote, never wake up from their 'theological slumber'[3] could easily have killed any interest in theology he may have had before he attended their lectures. When in later life he thought back on these days in Paris, he was to pour bitter scorn upon those who 'until the end of their lives will do nothing but hold disputations' and who 'in their *Summae* mix one thing with another and then mix them again, and who

---

[1] *Ratio studii ac legendi interpretandique aucthores*. For the date of its composition and its publications see P. S. Allen, *E.E.* vol. I, Introduction to no. 66, dated ⟨1497?⟩. *Ratio studii* is reprinted by A. Hyma = 'Erasmus and the Oxford Reformers (1493–1503)', *Nederlandsch Archief voor Kerkgeschiedenis*, N.S. vol. XXV (1932), pp. 129 ff. The revised edition published by Erasmus in 1512 is printed in *L.B.* vol. I, cols. 521 ff. The above quotation is: edit. of 1511, fol. e vij^v = Hyma, p. 130 = *L.B.* vol. I, col. 521 B.

[2] *Ratio studii*, edit. of 1511, fol. e viij^r = Hyma, p. 132 = *L.B.* vol. I, col. 522 B–C. This sentence influenced by A. Gellius, *Noctes Atticae*, XVI. 8. 17 is found in a similar form in Erasmus, *Enchiridion* (*L.B.* vol. V, col. 7 D) and *E.E.* vol. I, no. 108 (46–8), of ⟨October 1499⟩).

[3] *E.E.* vol. I, no. 64 (53–5), of ⟨August 1497⟩.

like quacks make and remake old out of new and new out of old, one of many, and many of one'.[1]

Yet, in spite of his aversion to scholasticism and the theology connected with its philosophy, he sometimes speaks in his letters of his occupation with, and inclination towards, theology. It has been asserted that Erasmus's letters written in Paris are not sincere and should therefore not be used as a source for his thought.[2] It is true, he was afraid that he might be recalled to the monastery of Steyn, especially as it was believed there that his life did not conform with that of a monk. It may therefore be assumed that his statements about his life and study were coloured to influence the recipients of his letters in his favour.

Erasmus's letters bear the mark of his time: they are full of excessive flattery but even in these phrases of eulogy his view sometimes comes out clearly, for example when he praises Henry of Bergen, the Bishop of Cambrai, as possessing both probity and erudition, 'the most beautiful combination in man but also the rarest'. For, he goes on, 'even virtue seems somewhat imperfect without erudition'.[3] This is a good example of the art of adulation by which Erasmus might retain the support of the Bishop, but it is also a dangerous statement revealing how greatly Erasmus valued the humanistic ideal of erudition. Erasmus probably felt that this praise might be counted against him as a monk, and he tried to retrieve the position by speaking against those who chose the heathen authors as their literary models instead of Christian writers, words which are in contradiction to other sentences in his letters of this period.[4] Continuing, he mentions Robert Gaguin, the French historian who, he writes, believed that 'religious subjects could become resplendent with the aid of secular riches if only purity of language is kept'. Erasmus continues, referring

---

[1] *E.E.* vol. III, no. 858 (38–42), of 14 August 1518. This letter is the preface to *Enchiridion Militis Christiani*.
[2] A. Hyma, 'Erasmus and the Oxford Reformers (1493–1503)', *loc. cit.* pp. 80ff.
[3] *E.E.* vol. I, no. 49 (72–82), of 7 November 1496.
[4] E.g. *E.E.* vol. I, no. 61 (131 ff.), of ⟨August 1497⟩. Cf. A. Hyma, *Nederlandsch Archief loc. cit.*

THE PHILOLOGICAL VIEW: ERASMUS OF ROTTERDAM

to St Augustine without mentioning his name, 'I would not disapprove of the use of the treasures of Egypt. But I do not recommend that the whole of Egypt be introduced.'[1] In these words we see Erasmus discussing the use of secular knowledge for theology. There is no doubt that this remark is sincere and serious, it is one of the keynotes for the understanding of Erasmus's correspondence in these years. It shows that Erasmus, like Gaguin, wished to make use of secular learning for religious subjects. The figurative mode of his speaking means, as in Augustine, that the knowledge of the pagans can be used to elucidate Holy Scripture.

If this thought was Erasmus's when he wrote this letter in 1496 it forced him to occupy himself in two different spheres: the study of the literature of antiquity and that of theology, albeit a theology different from the Parisian mode of teaching. If this is kept in mind it is easy to see that he could attack theology[2] as taught at the University of Paris where Scotist philosophy flourished, and at the same time praise theology as he understood it. In a sarcastic letter he attacked the theologians who follow the doctrine of Scotus and at the end he wrote:

Sweet Grey, do not mistake me. I would not have you construe this as directed against theology itself, which, as you know, I have always regarded with special reverence. I have only amused myself in making game of some pseudotheologians of our time, whose brains are rotten, their language barbarous, their intellects dull, their learning a bed of thorns, their manners rough, their life hypocritical, their talk full of venom, and their hearts as black as ink.[3]

Thomas Grey, an Englishman, had no power to send Erasmus back to the monastery of Steyn and yet Erasmus wrote: 'I have

---

[1] *E.E.* vol. I, no. 49 (92–6), of 7 November 1496. St Augustine, *De Doctrina Christiana*, 2, ch. 60 (*P.L.* vol. XXXIV, col. 63). Cf. *Enchiridion Militis Christiani* (*L.B.* vol. V, col. 25 F: '...perge...longius etiam in Gentilium litteris peregrinari, atque Ægyptias opes ad Dominici templi honestamentum convertere.' The sentence in Erasmus's letters is discussed by A. Hyma, *Nederlandsch Archief...*, p. 80, where a curious mistranslation occurs through the insertion of a negation in the first part of the first sentence.

[2] Cf. the personal attack: '...homo nihil me delectat; ipsa est theologia; imo scabies ipsa' (*E.E.* vol. I, no. 95 (43–5), of 2 May 1499).

[3] *E.E.* vol. I, no. 64 (86–92), of ⟨August 1497⟩ = N. vol. I, pp. 144–5.

always regarded Theology with special reverence.' Yet to the same Thomas Grey he had written about two months earlier that his work on literature 'has taught him not to yield to any of Fortune's storms'.[1] This phrase if taken seriously means that Erasmus considered literature (*literae*) as the forming power and as the most important influence in his life. It need not, however, imply the rejection of theology. The dual aim which Erasmus had in mind enabled him to praise both theology and literature without being insincere. It is, I suggest, natural that he considered the special predilections of the recipient of his letters and thus chose his words. He may, however, be reproached for using the word 'theology' without defining it. But it is very doubtful and even improbable that he could have explained the exact meaning of this word in any detail. As mentioned above, he had some knowledge of the early Fathers of the Church. He recognized that their theology was not dull and that their style was elegant. He saw them in complete contrast to the scholastic theology of his own time. He could employ their arguments against the modern theologians. He could, moreover, make use of the humanistic theory that the return to the literature of antiquity would produce a new flourishing civilization. It was the application of this thought to theology which made it necessary to return to the early Fathers who represented the great and brilliant period of Christian theology. A revival of their thought would restore theology to its 'pristine brightness', as he was to write in 1499. Apart from this he could follow their advice not to reject profane literature but to use pagan thought for theological studies. This was the help which they could give him. Yet he could not foresee the details of the work in which such a theology would involve him. Nor could he have a clear conception of the result of these studies. He saw an aim, the reconstitution of theology, but the way to this goal had not yet been paved. He was dissatisfied with the 'traditional view', he disliked the religiosity of the Brethren of the Common Life and the mysticism of the *Imitatio Christi*.

[1] E.E. vol. I, no. 58 (42–3), of ⟨July⟩ 1497 = N. vol. I, p. 138.

He was critical of the existing religious thoughts and communities. All this was negative, his only positive thought was that the philosophy and literature of antiquity could support Christianity, 'the best religion'.[1] This thought is humanistic with a turn characteristic of Erasmus's attempts to combine *literae* and religion. If his letters and his writings of this period are read with this thought in mind, they are seen to be generally consistent, in spite of certain contradictions, and, what is more important, they show the early stages of Erasmus's contribution to the history of thought—the use of the humanistic method for the interpretation of Holy Writ.

There were some fifteenth-century humanists who advocated the same principle. But with them this was only an idea occurring at some place in their works while Erasmus took it as the basis of his work and devoted a great part of his life to it. As mentioned above, there was among others the historian Gaguin with whose statements on the use of the 'treasures of Egypt' for religious subjects Erasmus agreed to a certain extent. Shortly before Erasmus wrote that letter about Gaguin, a similar thought had been expounded though not yet published by Jacob Wimpfeling in his *Isidoneus Germanicus*.[2] His reference to biblical studies is very careful and cautious:

Do we not learn from poets and orators how to speak Latin and how to write an ornate style? Do we not learn (from them) rhetorical flourishes, tropes and schemes which are often used in Holy Scripture?

Without the knowledge of poets and orators, he continues, we cannot understand the Fathers of the Church and the great Christian philosophers. If the reading of poets is forbidden, the

---

[1] *Antibarbari*, ed. Hyma, pp. 283–4=*L.B.* vol. x, col. 1713A; cf. *E.E.* vol. I, no. 49 (92–6), of 7 November 1496. Cf. R. Pfeiffer, 'Humanitas Erasmiana', *loc. cit.* p. 9.

[2] Erasmus's letter is of 7 November 1496, Wimpfeling's dedication is dated 21 June 1496. *Isidoneus Germanicus* is printed by Grüninger, without place or year. The year of printing has been suggested as 1497 (Ch. Schmidt, *Jean Grüninger 1483–1513, Répertoire Bibliographique Strasbourgeois jusque vers 1530*, vol. I (Strassburg, 1893), no. 33) and *c.* 1500 (*Catalogue of Books printed in the XVth Century now in the British Museum*, part I (1908), p. 116).

THE PHILOLOGICAL VIEW: ERASMUS OF ROTTERDAM

Fathers of the Church should not be read either since their works abound in quotations of poetry.[1]

This is a humanistic programme asserting that literary forms of rhetoric are of importance for the understanding of literary works and of theology. Compared with Erasmus's short note on Gaguin's view it is detailed, giving a definite line of approach. Unfortunately, there is no evidence of the opinion of Gaguin and Erasmus in this matter. In later times Erasmus was to mention the same points as Wimpfeling, but this does not necessarily imply his dependence on him.

Wimpfeling did not know Greek. He said that he had had no teacher for it 'in the bloom of his youth'. Thus Erasmus could not learn from him that it was necessary to go back to the original Greek of the New Testament. He could have known this from the writings of St Jerome and St Augustine. In the fifteenth century an important objection could be raised against such an attempt. Every text copied over and over again for centuries was bound to deteriorate owing to the negligence or ignorance of scribes or both. This had happened to the text of the Vulgate, as was well known. The Greek Bible, it was safe to assume, had suffered the same fate. Indeed it was easy to see that its text was inferior to that of the Vulgate. The Greeks, it could be argued, tried to prove their schismatic doctrines from their Greek texts. In order to do this, they must have purposely changed their Bible. Thus the Greek Bible could not be used for the correction of the Latin Vulgate—on the contrary, Greek was thought to be a hindrance for the interpretation of Holy Writ. The plea of the early Fathers of the Church was sound at their own time, but because of the schism their precepts could no longer be applied. This view on the Greek Bible, although expressed only in the sixteenth century, explains how new and unprecedented it was to use the knowledge of Greek for the exegesis of the Bible. It

---

[1] Pp. xvi^v, xxiii^v, xxiiii^r. On Wimpfeling's *Isidoneus* see Ch. Schmidt, *Histoire littéraire de l'Alsace à la fin du XVe et au commencement du XVIe siècle* (Paris, 1879), vol. I, pp. 22, 141 f.

elucidates why Erasmus did not have from the very beginning a clear idea of how to combine linguistic knowledge with theological work.[1]

It is true that twice in the fifteenth century the Latin and the Greek texts of parts of the Bible had been compared, but Erasmus had no knowledge of these works. One was Valla's *Adnotationes*, the importance of which for Erasmus will be discussed later, the other was Crastonus's edition of the Psalter, the first polyglot book of the Bible ever to be printed. Johannes Crastonus, the compiler of the Greek and Latin dictionary of about 1478, published this Psalter at Milan in 1481. The Greek wording and a Latin translation are printed side by side in two columns. The book of this Carmelite monk of Piacenza has a remarkable preface. Quoting as his authorities St Jerome and St Augustine he asserts the impossibility of reconstituting the Latin Bible without a comparison with the original Hebrew and Greek. As a result of such a comparison he claims to have found many corruptions of the Latin text caused either by mistakes of the scribes or, in most cases, by errors of the translators. As a result of this the text of Jerome cannot be recognized any more: 'I freely confess', he writes, 'that I do not know which is Jerome's translation.' After giving some examples of the corrupt text he continues: 'We have corrected the Latin text of the Psalms at about seventy places in accordance with the Greek truth. We have added those words which were missing.' Since Valla's *Adnotationes* were not published until 1505, Crastonus was the first openly to claim the superiority of the Greek text over the Latin translation. It is, of course, possible that he had learned this from reading a manuscript of Valla's work. Quoting Augustine for the justification of his view he attacks those who believed that 'Scripture is not subject to grammar'.

Erasmus would use the same arguments later, yet in all probability he did not know of Crastonus's edition of the Psalms. He

---

[1] Cf., for example, *E.E.* vol. II, no. 304 (86ff.), a letter from Dorp to Erasmus of ⟨c. September 1514⟩, and Erasmus's answer, *E.E.* vol. II, no. 337 (745ff.), of ⟨May *fin.*⟩ 1515.

does not mention his name. When, in 1501, he first started to compare the Latin and Greek texts of the Psalms, he seems to have been genuinely surprised at the discrepancies. It is likely that he would have mentioned the differences between the individual texts, had he had any knowledge of Crastonus's work.

In 1499, when he went to England, Erasmus's interests and thoughts were those of a humanist. He was convinced of the soundness and importance of the humanistic method which should prove useful for theology. But there is no evidence that he was sure how humanism could further theology. He might have considered theology from the literary and stylistic points of view only, believing that the representation of the theological subject was to be taught by the humanist. Some substance is given to this assumption by a remark in a letter accompanying some prayers which he had composed. These prayers, he believes, will cause the recipient 'to imbibe Christ together with the rudiments of erudition'.[1]

An example of Erasmus the humanist in all his charm and wit is found in a letter in which he relates how he broke up a religious disputation. He was present at a dinner at Oxford, perhaps at Magdalen College, in the last quarter of 1499. There was a discussion on the question of Cain's first sin. During this discussion a theologian argued with syllogistic, Erasmus with rhetorical arguments. 'At last,' Erasmus relates, 'when the dispute had continued rather long, and became more serious and solemn than was suitable to a banquet, I thought it time to take up my role of poet, and cheer the dinner with a more lively story, which might have the effect of breaking up the discussion.' With the permission of the others who participated in the discussion, Erasmus told a story recounting how Cain persuaded the angel who guarded Paradise to steal some of its grain. When God noticed this, He was angry with Cain for wishing to appease Him by his burnt-

[1] E.E. vol. I, no. 93 (96–106), of ⟨March?⟩ 1498/9. The date of this letter may well have been 1500 or 1501 (cf. P. S. Allen's Introduction to E.E. vol. I, no. 93).

offering of fruits; and so his sacrifice was not accepted.¹ This story, convincingly and artistically told, shows Erasmus the humanist, always ready to play the role of the poet. He had learned *bonae literae*, and used all his ability and all his learning to compose a story which has no religious content but is truly humanistic. He was, as he says in an understatement, 'little skilled in letters but a most warm admirer of them'.²

Conversations like these must have confirmed the impression made by Erasmus upon the learned circles in England that he was an accomplished humanist.³ John Colet, who was present at this dinner but who seems to have made Erasmus's acquaintance previously, had received certain information about Erasmus and wrote to him:

> When I was in Paris, Erasmus was not without celebrity in the mouth of the learned; an epistle of yours, addressed to Gaguin, in which you express your admiration of the labour and skill shown in his French History, served me, when I read it, as a sort of sample and taste of an accomplished man with the knowledge both of literature and of a multitude of other things.

This alludes to the only publication of Erasmus at that time, a letter printed in Gaguin's *De Origine et Gestis Francorum Compendium*.⁴ Yet to this praise of Erasmus's skill in letters Colet, giving expression to his own hopes, adds:

> But that which recommends you to me most is this, that the Reverend Father with whom you are staying, the Prior of the House and Church of Jesus Christ, affirmed to me yesterday, that in his judgement you were a singularly good man. Therefore, so far as learning and general knowledge and sincere goodness prevail with one, who rather seeks and wishes for these qualities than makes any profession of them, you, Erasmus, both are and ought to be most highly recommended to me.⁵

---

¹ *E.E.* vol. I, no. 116 esp. lines 32ff.=*N*. vol. I, pp. 216–19, of ⟨November 1499⟩.
² *E.E.* vol. I, no. 107 (41–2), to Colet, of ⟨October 1499⟩=*N*. vol. I, p. 207. Cf. the words found at the end of a letter to Colet on a theological subject: 'Vide, Colete, quam decorum obseruem, qui tam theologicam disputacionem poeticis fabulis claudam...' (*E.E.* vol. I, no. 109 (152–4), of ⟨October 1499⟩).
³ See, for example, *E.E.* vol. I, no. 114 (10–13), of 28 October 1499.
⁴ Pierre le Dru, Paris [not before 30 September 1495].
⁵ *E.E.* vol. I, no. 106 (8–14)= *N*. vol. I, pp. 205–6.

THE PHILOLOGICAL VIEW: ERASMUS OF ROTTERDAM

In spite of the flattery, there is sincerity in these words, and hope that this brilliant man is more than a flippant humanist.[1] This letter was the beginning of a friendship which was to last until Colet's death on 16 September 1519.

Colet,[2] the son of a rich merchant who had been Lord Mayor of London in 1486 and 1495, had studied in France and Italy. He was well versed in the seven liberal arts, and knew the works of Cicero, Plato and Plotinus, as well as the teachings of the Schoolmen. He had taken the degree of Master of Arts but not yet a degree in theology.[3] Yet he had, with great success, given lectures on St Paul 'publicly and gratuitously' at Oxford for three years. As Erasmus wrote later,

Though he had neither obtained nor sought for any degree in divinity, yet there was no doctor there, either of divinity or law, nor abbat, or other dignitary, but came to hear him...not being ashamed to learn, the old from the young, doctors from one who was no doctor.[4]

The attraction of these lectures consisted in the main in Colet's approach which was entirely different from that used by theologians at his time. Instead of indulging in 'mere subtleties and sophistical cavillings', which was in Erasmus's view the usual method of interpretation,[5] Colet expounded the text of St Paul, following in the main its own order. In this way Colet's words appealed directly to the public, for even while speaking on the

---

[1] Cf. the appreciation of Erasmus by Gaguin (*E.E.* vol. I, no. 44 (3-6), of 24 September ⟨1495?⟩).

[2] Colet's Biblical works, edited and translated by J. H. Lupton, are: *Two Treatises on the Hierarchies of Dionysius* (London, 1869); *An Exposition of St Paul's Epistle to the Romans* (London, 1873); *An Exposition of St Paul's First Epistle to the Corinthians* (London, 1874); *Letters to Radulphus on the Mosaic Account of the Creation, Together with Other Treatises* (London, 1876). Literature on Colet: F. Seebohm, *The Oxford Reformers John Colet, Erasmus, and Thomas More* (reprint, London, 1896); J. H. Lupton, *A Life of John Colet, D.D.* (2nd ed. London, 1909).

[3] *E.E.* vol. I, no. 181 (18 and note), to Colet ⟨*c.* December⟩ 1504.

[4] *E.E.* vol. IV, no. 1211 (244ff., esp. 285ff.), to Jodocus Jones of 13 June 1521. This letter contains bibliographical sketches of Vitrier and Colet. It has been translated by J. H Lupton, *The Lives of Jehan Vitrier...and John Colet* (London, 1883), from where the above translation is taken (pp. 23-4). Cf. *E.E.* vol. I, no. 108 (68-73) To Colet, of ⟨October 1499⟩.

[5] *E.E.* vol. I, no. 108 (21) = N. vol. I, p. 220. Cf. Erasmus, *Enchiridion* (*L.B.* vol. V, col. 8 D-E).

text only, his words were applied by the public to their own experiences. He avoided detailed word explanation and other digressions.

It is doubtful if Erasmus had heard Colet lecturing, but in his conversations with him dealing with religious subjects he must have become aware of his views. Both were 33 years old, and each was attracted by the other. Erasmus's letters and Colet's writings furnish considerable evidence that Colet profoundly influenced Erasmus.

Colet, as mentioned above, was well versed in the philosophy of the Schoolmen but when he visited France and Italy, 'he devoted himself entirely to the study of sacred authors'. These are words written by Erasmus in a biographical sketch of Colet, dated 1521. He continues that he was willing to read Scotus and Thomas Aquinas, although he preferred the early Fathers of the Church. Moreover, he was meticulous that felicitous expressions could be learned only through the study of the best authors. By nature he was, as he had confessed to Erasmus, inclined to 'incontinence, luxuriousness and indulgence in sleep; overmuch disposed to jests and raillery; and he was besides not wholly exempt from the taint of covetousness'. 'But these tendencies', Erasmus goes on, 'he combated so successfully by philosophy and sacred studies, by watching, fasting, and prayer, that he led the whole course of his life free from the pollutions of the world.' Colet's life was austere. Yet it is, I submit, characteristic of Erasmus to emphasize his human traits. 'Against his high temper', Erasmus writes, 'he contended with the help of reason, so as to brook admonition even from a servant.' At the end of the biographical sketch in which he also describes the life of Jehan Vitrier, Warden of the Franciscan Convent at St Omer, Erasmus compares these two men to whom he 'owes much', saying: 'In Colet were some traits which showed him to be but man. In Vitrier I never saw anything that savoured at all of human weakness.' Both were 'true and sincere Christians...'. 'If you take my word for it... you will not hesitate to enrol these two in the calendar of the saints, though no Pope should ever canonize them', Erasmus concludes.¹

Erasmus may have idealized Colet. But this biography is written with great warmth of feeling. Colet was a man, Erasmus believed, who lived sincerely as a Christian should, who was able to combine theology with the stylistic tendencies of humanism, who disliked the Schoolmen, who conversed with his friends about 'literature or about Christ'.² He was capable of interpreting the Bible in a way totally different from the syllogisms of the Schoolmen. Here then Erasmus saw a man who had solved his own difficulty: he was in orders subjecting his life to the precepts of religion and he was a humanist, he preferred the Fathers of the Church to the philosophical subtleties of the later exegesis, a man who did not tolerate 'a solecism or a barbaric impurity in language'.³ He had solved contradictions which had puzzled Erasmus and to which he had found no solution. Colet could show one way of combining humanistic and theological studies. 'When I hear my Colet, I seem to be listening to Plato himself', he wrote to Fisher at that time.⁴ I wonder if this sentence could not be explained by Erasmus's words in his *Enchiridion Militis Christiani*, published in 1503, where he mentions his preference for Platonists among philosophers because many of their sentences and especially the character of their speech come very near to those of the prophets and of the Gospels. This shows, I believe, Colet's influence on Erasmus. It was Colet's personality and the unity of his life which determined Erasmus to lay the stress of his own work not so much on secular literature as on studies which were connected with Holy Writ.

But such a resolve did not mean that Erasmus could go where Colet would lead him. They had many ideas in common,

[1] *E.E.* vol. IV, no. 1211, of 13 August 1521. Colet's biography is ll. 245-633, see esp. ll. 263-80, 519-27, 386-412, 618, 626-30. In Lupton's translation see pp. 21-3, 38-9, 30-2, 46-7.
[2] *E.E.* vol. IV, no. 1211 (323-5) = Lupton, p. 26.
[3] *E.E.* vol. IV, no. 1211 (329-30, 519-27).
[4] *E.E.* vol. I, no. 118 (21), of 5 December ⟨1499⟩ = N. vol. I, p. 226 (*Enchiridion*, L.B. vol. V, col. 7F).

the desire to fight against the Schoolmen, against the medieval grammarians and against abuses of the Church. All these views were directed against what they might have called the corruptions of their own time. But there were differences between them, great and important ones, which might easily have led to a parting of their ways if their friendship had been built on intellectual pursuits only. However, a personal affection bound them, so that Colet in his last letter to Erasmus in 1517 could write that he was half angry with him for sending greetings to him in letters addressed to others. 'I grieve', he wrote, 'to see you less regardful of me than of others.' Erasmus's feeling was expressed when on learning of Colet's death he wrote: 'Only half of me seems to be alive, now that Colet is dead', and 'For thirty years the news of nobody's death has distressed me so much as that of Colet.'[1]

There were differences between them from the very beginning about the interpretation of certain passages of the Bible when Erasmus defended the traditional view as held by most of the Fathers and by the Schoolmen whilst Colet favoured an exegesis based on one of St Jerome's passages. But there were dissensions between them of a graver nature, as can be learned from Colet's criticism of Thomas Aquinas. The reasons for Colet's rejection as narrated by Erasmus in 1521 were:

Without a full share of presumption, he [Thomas] never would have defined everything in that rash and overweening manner; and without something of a worldly spirit, he would not have so tainted the whole doctrine of Christ with his profane philosophy.[2]

These sentences, quoted as Colet's own words, reveal the essential differences between the two friends, differences not explicitly mentioned in their correspondence. The 'worldly spirit' of Thomas Aquinas can, in this connexion, only mean that he relied on his own knowledge and intellect to an extent which Colet thought excessive, for he believed that human understanding and

---

[1] *E.E.* vol. II, no. 593 (1-8); vol. IV, nos. 1025 (1), 1026 (1-4), 1030 (39-40), of 16 and 17 October 1519. Cf. J. H. Lupton, *Life of Dean Colet*, pp. 233-4.

[2] *E.E.* vol. IV, no. 1211 (425-44) = Lupton, pp. 32-3.

learning could lead only to profane philosophy, as distinct from the thought of the early Fathers of the Church like Dionysius the Areopagite, Origen, Cyprian and others.[1] In Colet's view 'the books of heathen authors savour of the Devil'. Pagan literature is of no use for the exegesis of the Bible. Holy Scripture, he maintains, can be understood 'by grace alone, and prayer, and by the help of Christ, and of faith'.

Now if any should say, as is often said, that to read heathen authors is of assistance for the right understanding of Holy Writ, let them reflect whether the very fact of such reliance being placed upon them, does not make them a chief obstacle to such understanding. For, in so acting, you distrust your power of understanding the Scriptures by grace alone, and prayer, and by the help of Christ, and of faith; but think you can do so through the means and assistance of the heathens.... Those books alone ought to be read in which there is a salutary flavour of Christ; in which Christ is set forth for us to feast upon. Those books in which Christ is not found, are but a table of devils. Do not become readers of philosophers, companions of devils. In the choice and well-stored table of Holy Scripture all things are contained that belong to the truth. And doubt not that the mind which craves for anything to feed on beyond the truth, is in an unhealthy state, and is devoid of Christ. The truth, moreover, is understood by grace; grace is procured by our prayers being heard; our prayers are heard, when whetted by devotion and strengthened by fasting. To have recourse to other means is mere infatuation.[2]

This view is in full harmony with his saying that 'an interpreter of Scripture is not called upon to play the part of the grammarian, or to examine words overminutely'.[3] It is necessary to see this remark against the background of scholastic exegesis to which Colet was opposed. It seems that Colet never changed his view that 'by grace alone' God's word can be understood. Not until

---

[1] *E.E.* vol. IV, no. 1211 (270–3)=Lupton, pp. 21–2.

[2] *An Exposition of St Paul's First Epistle to the Corinthians* ed. Lupton, pp. 110–11, 238–9, quoted by J. H. Lupton, *A Life of John Colet*, p. 76, and by A. Hyma, 'Erasmus and the Oxford Reformers, 1493–1503', *Nederlandsch Archief...*, pp. 99 ff. and *idem*, 'Erasmus and the Oxford Reformers, 1493–1503', *Bijdragen voor Vaderlandsche Geschiedenis en Oudheidkunde*, vol. VII, 7 (1936), p. 142.

[3] *Letters to Radulphus*, pp. 81 and 222.

he had received Erasmus's New Testament in 1516 was he to deplore his ignorance of Greek. Only then did he understand the importance of languages for the exegesis of Holy Writ. Colet's letter to Erasmus which has been quoted in the chapter on Reuchlin bears witness to the consistency of his thought:

Ah, Erasmus, of books and of knowledge there is no end; but there is nothing better for this short term of ours, than that we should live a pure and holy life, and daily do our best to be cleansed and enlightened, and perfected. This is promised by those Pythagorean and cabbalistic ideas of Reuchlin: it will in my judgement never be attained, but by the ardent love and imitation of Jesus. Wherefore it is my earnest wish, that leaving all indirect sources, we may proceed by a short method of truth. I wish for this with all my strength.[1]

These views are, it seems, in complete contradiction to all the ideas of Erasmus. Indeed, it is scarcely feasible to believe in the friendship of these two men if Erasmus is considered to be only a man of enlightenment comparable with Voltaire.

It is very difficult to discover the extent of the influence Colet had on Erasmus. There are students of Erasmus who deny it altogether, others who believe it to have been very strong, while some consider it to have been 'plus morale qu'intellectuelle'.[2] The following pages dealing with Erasmus's development during his stay in England will show how difficult it is to assess the extent to which this influence may have shaped his thought.

Among Erasmus's letters there is an interesting one dating from this period, in which he refuses to work together with Colet. From it it may be concluded that Colet had invited him to lecture at Oxford, and to give a continuous commentary on Moses or Isaiah as Colet had done on St Paul. It may possibly be that as a second choice Colet suggested lectures on poetry and rhetoric.[3]

---

[1] *E.E.* vol. II, no. 593 (15–21) = *N.* vol. II, pp. 596–7 (where the last sentence is missing). Some changes in the translation are due to Allen's amendment of the text.

[2] J. B. Pineau, *Érasme, sa pensée religieuse* (Paris, 1924), p. 90 n. 74.

[3] *E.E.* vol. I, no. 108 (77–9 and 95–6), of ⟨October 1499⟩. P. S. Allen, Introduction to *E.E.* vol. I, no. 108, assumes that this letter was written at the beginning of Erasmus's stay at Oxford, while Nichols in his *Translation* of Erasmus's letters (vol. I, pp. 220–3) dates it later. At first sight it seems that it was composed shortly before Erasmus left Oxford. But Allen's arguments carry weight and may well be right.

THE PHILOLOGICAL VIEW: ERASMUS OF ROTTERDAM

This invitation forced Erasmus to choose between these two disciplines: theology or the arts. This letter of 1499 containing his refusal of Colet's offer gives a reasoned account of Erasmus's position at this time, namely that of a humanist who holds theology in high regard. It is in agreement with his earlier writings, but is more precise in describing the special character which the future theology should have.

Like Colet he hates 'that sort of neoteric divines who grow old in mere subtleties and sophistical cavillings'.[1] The reason for his attitude is that these sophisms do not make men wise, but conceited and contentious dabblers. As a result of these studies,

Theology, the queen of all disciplines, enriched and adorned by the eloquence of the ancient writers, has lost her grace through the stammering and sordidness of most impure language.[2]

They confuse everything while they try to solve everything.

Thus theology, once most august and full of majesty, is now almost mute, destitute and ragged.[3]

Some of the theological questions are scarcely fit to be listened to by religious people, e.g. whether God could assume the nature of the devil or of an ass. In the discussion of such questions 'we grow old and die', he says bitterly; for these, he continues, we despise all *literae*. He adds:

In our day, theology, which ought to be at the head of all *literae*, is mainly studied by persons who from their dulness and lack of sense are scarcely fit for any *literae* at all.[4]

In this part of the letter Erasmus brings out the points of agreement between himself and Colet in an argument which is entirely

---

[1] *E.E.* vol. I, no. 108 (20–1) = N. vol. I, p. 220. Nichols leaves out ll. 22–48 (*E.E.*) in his translation without any indication of this omission. A more complete translation of this letter is found in F. Seebohm, *Oxford Reformers*, pp. 129–33, and a complete version is found in J. J. Mangan, *Life, Character and Influence of Desiderius Erasmus of Rotterdam*, vol. I (London, 1928), pp. 114–17. (The only omission of a few words is noted on p. 115.)
[2] *E.E.* vol. I, no. 108 (28–31); cf. *ibid.* ll. 52–6.
[3] *E.E.* vol. I, no. 108 (33–5).
[4] *E.E.* vol. I, no. 108 (41–51) = N. vol. I, p. 220.

humanistic. In words which remind the reader of his *Ratio Studii* he attacks 'the neoteric divines who grow old in mere subtleties and sophisticated cavillings'.[1] As a contrast he conjures up the splendour of the theology of old against which the neoteric divines fight. Colet is the hero who will bring about 'the restoration of genuine theology to its pristine brightness and dignity'[2] and who courageously withstands the onslaught of the enemies. This part of the letter, in spite of its flattery, contains at least two truthful elements: Erasmus sees that the new religious studies as envisaged by him will be a restoration of theology which is to be achieved through avoidance of sophisms of every kind and through eloquence and purity of language. Secondly, Colet fights against scholastic theology and therefore he has 'undertaken a pious work as regards theology itself'. Erasmus does not speak explicitly of Colet's attitude to the Fathers of the Church. But in another letter of the same time, not addressed to Colet, he calls him 'asserter and champion of the old theology'.[3] This can only mean that he favoured the theology of the Fathers of the Church, which was 'most august and full of majesty'.[4] This high estimate of the Fathers completely tallies with Erasmus's views as expressed in his writings and letters before his arrival in England. His argument, as mentioned above, depended upon Valla's theory that the decay of language is the reason for the decline of civilization. This thought is applied to the decay of that theology which is not based on those old and elegant *literae*[5] and which has thus lost its power. This leads logically to the first argument stated above: a return to a pure style must, of necessity, restore theology.

In the second part of this letter Erasmus speaks of his reasons for refusing Colet's offer. In spite of its length this passage may be quoted in full.

I do not wonder at your taking such a burden on your shoulders, for you are equal to it. I do wonder at your inviting so insignificant a

---

[1] *E.E.* vol. I, no. 108 (20–1, 46–7). [2] *E.E.* vol. I, no. 108 (57–8).
[3] *E.E.* vol. I, no. 116 (12–13), of ⟨November 1499⟩ to John Sixtin=*N.* vol. I, p. 215.
[4] *E.E.* vol. I, no. 108 (33–5). [5] *E.E.* vol. I, no. 108 (24).

person as me to be partner in so noble a task. You exhort me, or rather you urge me with reproaches, to endeavour to kindle the studies of this university [Oxford]—sunk in torpor as you write, during these winter months—by commenting on the ancient Moses or the eloquent Isaiah, in the same way as you have done on St Paul. But I, who have learned to live with myself, and know how scanty my equipment is, can neither claim the learning required for such a task, nor do I think that I possess the strength of mind to sustain the jealousy of so many men, who would be eager to maintain their own ground. The campaign is one that demands, not a tiro, but a practised general. Neither should you call me immodest in declining a position which it would be most immodest for me to accept. You are not acting wisely, Colet, in demanding water from a pumice-stone, as Plautus says. With what effrontery shall I teach what I have never learned? How am I to warm the coldness of others, while I am shivering myself? I should deem myself more rash than rashness itself if I tried my strength at present in so great an enterprise, and, according to the Greek proverb, tried to run before I could walk.

But you say you expected of me some work of this kind, and complain that you have been disappointed. In that case you must find fault with yourself, not with me. I have not disappointed you, for I never either promised or held out any prospect of such a thing. It is you that have deceived yourself, by not believing what I said truly of my character. Neither again did I come here to teach poetry or rhetoric. These studies ceased to be agreeable to me when they ceased to be necessary. I decline this task, because it is below my purpose, as I do the other, as it is above my strength. As to the one your reproach, my dear Colet, is undeserved, because I never proposed to myself the profession of what is called *secular literae*; and to the other you exhort me in vain, because I am conscious of my own unfitness for it. And if I were ever so fit, it could not be, as I am returning before long to Paris. In the meantime...I betook myself to this learned university [Oxford], to spend a month or two with men like you, rather than with those gold-chained courtiers.

However, I am so far from opposing your glorious and sacred endeavours, that, not being yet a suitable fellow-labourer, I will promise my earnest encouragement and sympathy. And further when I am conscious of the needful strength, I will put myself on your side, and will make an earnest, if not a successful, effort in defence of

Theology. Meantime nothing could be more delightful for me than to discuss daily between ourselves, either by word of mouth or by letter, some subject of sacred literature.[1]

It must be asked again whether this part of the letter is sincere or whether the excuses are mere rhetorical phrases couched in the form of assertion of modesty. The phrasing of some sentences may be extravagant. But again the reader finds agreements with earlier letters. This does not necessarily prove sincerity, for humanistic reasoning was natural to Erasmus and he therefore used it for the refusal of Colet's offer. Yet this argument is strongly connected with his promise to come to Colet's help at some future time. However vague this promise may sound, it was fulfilled by Erasmus in his later work on the Bible.

It will be remembered that in a letter of about June 1489 Erasmus had traced barbarism as resulting from the teaching of illiterate people who had learned nothing and taught what they did not know.[2] This thought was, I believe, foremost in Erasmus's mind when writing to Colet: 'With what effrontery shall I teach what I have never learned?' Since his erudition was not sufficient, he would act like a 'barbarian' if he taught theology. He had first to learn the subject before he could put himself on Colet's side for its defence. He expressly states that he cannot 'claim the learning required for such a task'. Obviously he has to learn theology as he understands it, something, it seems, to which his study of *literae* is preparatory, but which is different from it.

It has been shown above that Erasmus sometimes made statements which amaze the reader since he stresses *literae* or theology to the detriment of each other. In his letter of 1499 he again refers to this question placing himself between these two disciplines: poetry and rhetoric, which he also calls 'secular *literae*', are below his purpose; theology, the exposition of a commentary on Moses or Isaiah, is above his strength. In these words he places the arts

---

[1] *E.E.* vol. I, no. 108 (74–114) = *N.* vol. I, pp. 221–3 (with some changes of Nichol's translation).

[2] *E.E.* vol. I, no. 23 (78–100); see above p. 97.

lower than theology, 'the queen of all disciplines', which, as he pointed out at the beginning of the letter, should be 'built on *literae* more ancient and more elegant'. Erasmus, it seems probable, was still brooding about the question which he had discussed with Gaguin in 1496, namely to what extent the knowledge of pagan literary works should be used for the interpretation of theological subjects. He now recognized, or at any rate asserted, that he had neither sufficient learning for the achievement of Colet's quest, nor the strength necessary to resist all the attacks which such a work would invite. Erasmus does not define what he means when speaking about his lack of learning, and it seems impossible to find out its exact significance since neither are all the letters which were exchanged between them preserved, nor do we know the contents of their conversations to which Erasmus expressly refers.[1] He only contrasts poetry and rhetoric with theology—and maintains of poetry and rhetoric that 'these studies ceased to be agreeable to me when they ceased to be necessary'. Since when did Erasmus hold this view? Is this sentence nothing but rhetoric, to explain his position between two different fields of learning? I do not think so, for five years later he still uses the same phrase, adding one word which limits its meaning. He was to ask Colet 'to rescue him from *that kind* of literary work which has ceased to be agreeable to him', only to continue, that he wishes to prepare a second edition of his *Adagia*.[2] This reference, although explaining the earlier sentence to some extent, does not guarantee the identity of Erasmus's view over these five years. As it is, the words of 1499, lacking as they do a definition of *literae*, cannot be clearly understood. But if this sentence is to be brought into line with the first part of the letter, where stress is laid on the importance of style for theology, it may be assumed that the necessity of learning poetry and rhetoric had disappeared for him who had spent a considerable time in learning 'secular *literae*'. This thought, if this

---

[1] *E.E.* vol. I, no. 108 (92ff., 112–13).
[2] *E.E.* vol. I, no. 181 (77–8), of ⟨*c.* December⟩ 1504: '...quoad potes adiuues, atque ab *iis* literis, quae mihi iam dulces esse desierunt.'

interpretation could be accepted, was to be expressed again in his *Enchiridion*.[1] Even Colet might have agreed to it, for he too demanded purity of style.

The letter ends with Erasmus's plan to go back to Paris and with his promise to work with Colet at some future date. It seems that Colet's influence on Erasmus can be described in the following terms: he had shown him the possibility of an exegesis free from the sophisms of modern expositors of the Bible written in a style worthy of praise;[2] Erasmus called him, in a letter not addressed to him, 'an asserter and champion of the old theology'.[3] What would be the exegesis as advocated by Colet? It would be theological, not philological without over-minute regard to the single words. The text to be used would be the Vulgate, although once, but only once, referring to Origen, Jerome and other commentators, Colet protests that the words of Moses could be understood only if the original Hebrew were consulted.[4] But his own interpretation is based on the Vulgate. It was only in 1516, after he had obtained Erasmus's New Testament, that he decided he should learn Greek 'without which we are nothing',[5] as he wrote. There is another indication that Colet did not see the importance of the original Hebrew or Greek texts. He asked Erasmus to comment on parts of the Old Testament, although Erasmus had no knowledge of Hebrew. He would scarcely have done so if he had fully understood the essential significance of the Hebrew Bible. It is probable, as will be seen below, that Colet approved of a new translation of the New Testament which, he may have assumed, would help theological exegesis. But this was a far cry from learning Greek because of the belief that only the original text of the Bible could serve as a basis for interpretation.

It is, I think, very unlikely that in 1499 Erasmus's wish to learn Greek was connected with biblical studies. To learn Greek was, of course, the desire of many a humanist at that time. As mentioned

---

[1] *L.B.* vol. v, col. 7D–E. [2] *E.E.* vol. I, no. 107 ⟨57–66⟩, of ⟨October 1499⟩.
[3] *E.E.* vol. I, no. 116 (12–13), to John Sixtin, of ⟨November 1499⟩.
[4] *Letters to Radulphus*, etc., pp. 3, 167.
[5] *E.E.* vol. II, no. 423 (13–15), of 20 June ⟨1516⟩.

THE PHILOLOGICAL VIEW: ERASMUS OF ROTTERDAM

above, the study of Greek does not of necessity lead to theological studies but rather to secular classical literature. When in 1499 he was faced with Colet's request to work with him, was Erasmus fully aware that Scripture 'must be studied in the original form or the nearest form to that which critical scholarship could discover'?[1] I do not think so. There is no evidence that while in England he fully comprehended the significance of Greek for theological studies. We are able to observe how, in the following years, this thought gradually ripened until he understood that Greek is the necessary prerequisite for the restoration of the theology. And how could it be different? For how could he judge of the value of Greek before he knew the language and before he himself had found out the discrepancies between the Latin text of the Vulgate and the Greek Bible? Nobody was able to assist him in arriving at this idea, which was contrary to all tradition.

The question which seems to have occupied Colet and Erasmus was not the tradition of the Bible but rather the restoration of exegesis. It is known that they discussed the question whether the sentences of Holy Scripture have one sense only, as Colet thought, or many, as Erasmus asserted.[2] They wished to return to the interpretation of the early Fathers of the Church and to abandon the sophisms of modern expositors. Colet had shown one way to attempt this. The importance of his example cannot be overestimated, even if Erasmus had to find another method for the achievement of this result.

Erasmus learned Greek after his return from England at the beginning of 1500 and he prepared an edition of *Adagia*,[3] a work

---

[1] For a different interpretation see P. S. Allen, *Erasmus, Lectures and Wayfaring Studies* (Oxford, 1934), p. 42.

[2] L.B. vol. v, cols. 1267A, 1292C ff. For the history of this text see P. S. Allen, Introduction to E.E. vol. I, no. 108 (245–6).

[3] The *Adagiorum Collectanea* were first published by Jo. Philippus, Paris, in 1500. The *Preface* written in ⟨June 1500⟩ is reprinted in E.E. vol. I, no. 126. For changes of this edition and for later reprints see P. S. Allen's Introductions to nos. 126, 211, 269, 1204, 1659, 2022, 2023, 2773, 3092. See Allen's Introduction to E.E. vol. IV, no. 1175 about the relationship between Erasmus's *Adagia* and Polydore Vergil's *Adagiorum Liber*. For editions see B.Er.2.

the contents of which seem to be leading far away from his promise to Colet. Yet this work represents a new stage in the development of Erasmus's attitude towards the Bible. It reveals that Erasmus could not give up his humanistic ideas and that poetic and rhetoric had not 'ceased to be agreeable' to him, for he writes in the Preface: 'There is no sweetness in life without *literae*.' Using a metaphor which he was to repeat in a letter to Colet at the end of 1504 in a more austere form, he says:

> Laying aside all serious labours, and indulging in a more dainty kind of study, I strolled through the gardens provided by various authors, culling as I went adages most remarkable for their antiquity and excellence, like so many flowers of various sorts, of which I have made a nosegay.[1]

In this preface, which contains a defence of humanistic studies, he mentions the usefulness of a collection of proverbs. Proverbs, he asserts, adorn ordinary speech, and their knowledge is therefore essential for writing a good style.[2] To those, however, who object to every kind of rhetoric Erasmus answers that rhetorical devices and adages are used by everybody, by writers of antiquity, by the Fathers of the Church, and by Italian humanists, not only by 'orators but by the wise, by prophets and theologians', for they are enigmatic and mysterious. They contain 'something divine, that is accommodated to celestial things'. Proverbs therefore are not merely rhetorical varnish, but they are found, for instance, in the Gospels and in the Epistles of St Paul. Thus it is essential for the understanding of both secular writing and Holy Scripture to learn the meaning of adages.[3]

It is true that before Erasmus Wimpfeling had mentioned that rhetorical figures are found in poetry and in Holy Writ. Erasmus's Preface excels in its presentation, its style, vividness and clarity of thought. He insists that his collection of proverbs proves the value of literary form for the comprehension of both literary

---
[1] *E.E.* vol. I, no. 126 (8–19), of ⟨June 1500⟩=*N.* vol. I, p. 243; cf. *E.E.* vol. I, no. 181 (89–90).
[2] *E.E.* vol. I, no. 126 (32 ff.).   [3] *E.E.* vol. I, no. 126 (53–197).

THE PHILOLOGICAL VIEW: ERASMUS OF ROTTERDAM

works and Holy Writ.[1] This is a new and important thought which, expressed in an abstract way, asserts that from the literary point of view letters, be they sacred or secular, are identical and can therefore be subjected to the same method of literary criticism for their understanding. Thus Erasmus exceeds by far the demand of humanists like Wimpfeling whose thought was concerned only with rhetorical figures of speech. Proverbs, Erasmus states, can contain 'the deepest secrets', as can be seen in the Book of Proverbs.[2] In such words we can discover the first expressions of a thought which was to be of the highest importance for Erasmus. The mysteries of God's word are revealed in certain forms known to man as proverbs. The knowledge of this form, the understanding of proverbs, will therefore lead to the comprehension of Holy Writ. Erasmus was now able to unite the study of poets and orators with that of the Bible. Everything he could learn from profane literature could usefully be transferred to the literary interpretation of the Bible. In defence of his collections of adages Erasmus had stated one of the guiding principles of that method which he was never to forsake. He had found the humanistic approach to the understanding of some of the mysteries of God's word.

When making his collection of *Adagia* Erasmus's knowledge of Greek was very limited. In 1504 he even wrote that he was sorry to have published the edition because of his ignorance of Greek.[3] In December 1500, while learning this language, he may have become more conscious of its importance for biblical studies, especially under the influence of St Jerome's works 'upon whom I am preparing commentaries.... It is incredible, how my heart burns to bring all my poor lucubrations to completion, at the same time to attain some moderate capacity in Greek, and then to devote myself entirely to the study of sacred literature, as for some

---

[1] *E.E.* vol. I, no. 126 (106–9, 172–97).
[2] *E.E.* vol. I, no. 126 (179–81) '...libri, quem Salomonis prouerbia nominant, altissimis mysteriis referti....'
[3] *E.E.* vol. I, no. 181 (85–6). For Erasmus's learning of Greek see P. S. Allen, *E.E.* vol. I, Appendix VI, p. 592.

time I have longed to do.' Continuing, he says that within the next three years ('if I am permitted to live three years') he will be able to excel by his 'treatment of theological literature'.[1] Erasmus envisages a further delay before he can work on theological subjects. It may be, though it is in no way certain, that by implication he intended to convey the impression that he wished to use the knowledge of Greek for the 'treatment of theological literature'. In this case this letter would be the earliest evidence to this effect.[2]

In the Preface to his *Adagia* he praises St Jerome as the best theologian on account of his erudition and his style. A short time later at the end of the year 1500 he states that he intends to correct the text of, and to write a commentary on, St Jerome's letters.

> Good Heavens! [he exclaims], shall the names of Scotus, Albertus, and writers still less polished be shouted in all the schools, and that singular champion, exponent and light of our religion, who deserves to be the one person celebrated—shall he be the only one of whom nothing is said?

But this state of affairs will be changed, for through his new commentary 'the glory of Jerome will shine forth with a new light'.[3] When a few months later he again speaks of his work on St Jerome, he stresses the difficulties facing the editor, much more than before. He understands how arduous a task it is to restore St Jerome's works which are, owing to ignorance, mutilated and full of mistakes. St Jerome is, he maintains, the only one of the ecclesiastics who acquired both sacred and secular learning.[4] The

---

[1] *E.E.* vol. I, no. 138 (37–58)=N. vol. I, pp. 283–4, to James Batt, of 11 December ⟨1500⟩. P. S. Allen refers in a note to this letter to Erasmus's promise to Colet. A. Hyma, 'Erasmus and the Oxford Reformers', *Nederlandsch Archief*, *loc. cit.* pp. 120–1, interprets these lines: 'Only he (Erasmus) had talked that way many times before and had never done what he had promised to do, that is to "devote himself entirely to the study of Sacred Literature".'

[2] Allen's interpretation (*E.E.* vol. II, Appendix VI, p. 592 and n. 26) is more confident: Erasmus is 'convinced that Greek is necessary to all serious students, especially in Theology'.

[3] *E.E.* vol. I, no. 141 (16–49)=N. vol. I, p. 289, of 18 December ⟨1500⟩.

[4] *E.E.* vol. I, no. 149 (56–65), of ⟨16 March ? 1501⟩. Cf. *Antibarbari*, ed. Hyma, p. 324=L.B. vol. x, col. 1739 D–E.

THE PHILOLOGICAL VIEW: ERASMUS OF ROTTERDAM

humanist Erasmus has found the one of the Latin Fathers who had conceptions similar to his own views. Having made up his mind that St Jerome should be his guide he writes in the same letter:

For my own part, I choose to follow the path to which St Jerome, with the noble band of so many ancient fathers, invites us. I had rather, so help me Heaven, lose my senses with them, than be as wise as you please with the herd of neoteric divines.[1]

This is a real programme, a definite plan of work to which he will cling with tenacity and vigour for many years to come. He cannot when writing this letter see all the pitfalls and difficulties which he will have to overcome. He cannot foresee that he will not publish Jerome's letters until 1516. But during all these years he will never falter. In the dedicatory letter of this edition of 1516 and in the prefaces to the several volumes all the points mentioned in 1500 and 1501 are found: Jerome, the stylist who is equal to Cicero and who has knowledge of sacred and profane *literae* superior to everybody else, is the greatest among the Latin Fathers, 'his old theology' is alone worthy of the name 'theology'.[2]

When he started his work on Jerome in 1500, Erasmus was well aware that his aim was a complete evaluation of Jerome's importance for Christianity. This, he was sure, would be a great step towards 'the restoration of that old genuine theology...to its pristine brightness and dignity' and it would be a decisive blow against the 'neoteric divines'. These words are, however, taken from the letter Erasmus wrote to Colet in 1499, when he promised him that he would defend theology on Colet's side. Colet did not then expect that Erasmus would edit a representative of

---

[1] *E.E.* vol. I, no. 149 (52–65) = N. vol. I, p. 314. Cf. *E.E.* vol. I, no. 161 (35–42), of 18 July ⟨1501⟩.

[2] The dedicatory letter, dated 1 April 1516, is reprinted in *E.E.* vol. II, no. 396, where in the Introduction Allen mentions those letters which throw light on Erasmus's attitude on Jerome (p. 210). The Prefaces of the edition of 1516 (Froben, Basel) are important to see the continuity of Erasmus's views, e.g. vol. III, fol. 1ᵛ (last lines) and esp. vol. II, fol. 2ʳ. Cf. *E.E.* vol. II, no. 334 (99–162), of ⟨15 May⟩ 1515. A defence of Jerome is found *E.E.* vol. III, no. 844 of 15 May 1518.

the 'old theology'. Yet this was one way in which Erasmus attempted to bring about a restoration of theological thought.[1]

While working on St Jerome in 1501, Erasmus compared the Greek text of some Psalms with the Latin wording of the Vulgate. While doing so he discovered discrepancies between the texts in the different languages. These differences between the two texts were caused, he noticed, by the literal method of the translation into Latin. The Latin text, he concluded, could not be understood unless it was compared with the Greek from which it had been rendered. As examples he quoted Psalm l. 5 (li. 3) and Psalm xci. 15 (xcii. 14).[2] This important advance in Erasmus's scholarship manifests a further stage in his development. In 1501, Erasmus became aware, as Reuchlin had done in 1488, of inaccuracies in the Latin text of the Bible and perceived that it was necessary to consult the wording of the language from which in his view the Psalms had been translated. Erasmus's position in 1501 is as follows: He points out that the misunderstanding of the Latin text of Psalm l. 5 (li. 3) leads to a theological exegesis of the verse which is wrong. He therefore concludes that it is impossible to use the Latin wording as a basis for a theological interpretation. Owing to the literal rendering from Greek the Latin can only be understood if it is compared with the Greek. It is therefore neces-

[1] E.E. vol. I, no. 108 (56–8 and 109–11); no. 138 (47); no. 139 (42–5), of ⟨c. 12 December⟩ 1500.

[2] It seems that the examples chosen by Erasmus for the elucidation of the differences were not taken from earlier writers. Here is as much evidence as I have collected: Erasmus (E.E. vol. I, no. 149 (26–42)) quotes the Greek text of part of Psalm l. 5 (li. 3) together with its translation in the Vulgate, and one Greek word with its Latin rendering of Psalm xci. 15 (xcii. 14). The Greek texts of all the printed editions before 1501 are identical but in Psalm xci. 15 (xcii. 14) Erasmus quotes the accusative instead of the nominative case (the printed edition of 1486 has a misprint εὐπαθοῦντεις for εὐπαθοῦντες). In Psalm l. 5 (li. 3) Erasmus criticizes the rendering of ἐνώπιον with contra. St Jerome (Liber Psalmorum iuxta Septuaginta interpretes (P.L. vol. XXIX, cols 207–8)) translated this Greek word with coram which is also found in Augustine's text (Enarratio in Psalmos (P.L. vol. XXXVI, col. 589)) and in Crastonus's edition of 1481. Erasmus explains the Greek word 'cum...situm significet...e regione, id est in conspectu'. As to Psalm xci. 15 (xcii. 14) St Jerome (P.L. vol. XXIX, cols. 291–2) and Crastonus have the syllable-for-syllable translation: εὐπαθοῦντες = bene patientes. Augustine (P.L. vol. XXXVI, col. 1180) renders 'tranquilli'. Erasmus criticizes bene patientes for being an expression which cannot be understood in Latin.

THE PHILOLOGICAL VIEW: ERASMUS OF ROTTERDAM

sary for the theologian to have a thorough knowledge of Greek. This conclusion, however, was, as pointed out above, a dangerous conception. He therefore was sure that he would be attacked because of his use of Greek, a language which could be a formidable weapon against the 'neoteric divines'. Seeing this he at once prepared his defences. In the same letter he points out that his undertaking is in conformity with the Decreta of the Church, for at the Council of Vienne of 1311/12 it was promulgated 'that there should be provided in the chief Academies...persons capable of teaching perfectly the Hebrew, Greek, and Latin languages, inasmuch as they held that without this knowledge sacred literature could not be apprehended, still less discussed'. The reference to this Council is one of his main defences during the whole of his life, enabling him to say as in the above letter: '...I have on my side the sacred authority of the Pontifical Council....'[1]

This letter of 1501, which is also one of the testimonies about his work on St Jerome, proves that Erasmus has found the basis for his future work. This discovery forces him to define more clearly than before the purpose and the aim he wishes to set for himself. The comparison of the Greek and Latin texts of the Psalms had shown him that the theological exegesis 'treating divine mysteries' depends on a humble task, namely that of understanding the words of the biblical text:

We have in Latin at best some small streams and turbid pools,[2] while they [the Greeks] have the clearest springs and rivers flowing with gold.[3]

---

[1] *E.E.* vol. I, no. 149 (42–51) = *N.* vol. I, pp. 313–14, of ⟨16 March ? 1501⟩. The reason for the introduction of languages at the universities was the recognition that they are useful for the conversion of the infidels. It is doubtful if Greek was included in these languages. See *E.E.* vol. I, no. 182 (181 n.).

[2] Cf. Reuchlin's sentence of 1512 quoted above.

[3] Cf. Erasmus's *Ratio* found in Erasmus's New Testament of 1519 (p. 20) where the reader is asked to compare the old theologians, e.g. Jerome ('illic') with the modern divines ('hic') in the following words: 'Videbit illic aureum quoddam ire flumen: hic tenues quosdam rivulos, eosque nec puros admodum, nec suo fonti respondentes. Illic velut in felicissimis hortis affatim tum oblectaberis, tum expleberis, dum hic inter spineta sterilia dilaceraris, ac torqueris....' Cf. the preface to the second volume of Erasmus's edition of Jerome's works of 1516 (Fo. 2^{r-v}): 'reliquos (theologos) ut limpidos rivos admireris. hic (Hieronymus) ceu dives quoddam et aureum flumen, universas opes secum volvit et defert.'

THE PHILOLOGICAL VIEW: ERASMUS OF ROTTERDAM

I see it is the merest madness to touch with the little finger that principal part of theology, which treats of divine mysteries, without being furnished with the apparatus of Greek, when those who have translated the sacred books have in their scrupulous interpretation so rendered the Greek phrases that not even that primary meaning which our theologians call 'literal' can be perceived by those who are not Greek scholars.[1]

The clearness of this letter is, however, deceiving. As so often, Erasmus is inconsistent in his views. The examples given in this letter have forced him to certain conclusions but he is not yet willing or able to make general rules for the study of the Bible. His thought about the relative value of the primary and mystical meanings was influenced by tradition but he did not yet perceive the full significance of the literal meaning. At about the same time he wrote his *Enchiridion Militis Christiani*,[2] in which he suggests that the literal exegesis of the Bible should be neglected. In this book sentences are found which are in agreement with Colet's opinion that words should not be examined too closely. As in Colet's writings this precept is directed against the exegesis of the Schoolmen who, in his view, interpreted the single words without any perception of the meaning:

Of the interpretours of scripture chose them aboue al other that go farthest from the lettre which chefely next after Paule be Origene, Ambrose, Jerom and Augustyne. For I se the diuines of later tyme stycke very moche in the lettre and with good wyll gyue more study to subtyle and deceytfull argumentes than to serche out the mysteryes as though Paule hath not sayd truly our lawe to be spirituall.

---

[1] *E.E.* vol. I, no. 149 (19–26) = *N.* vol. I, p. 313.

[2] For editions see *B.Er.* 1. For the date see P. S. Allen, Introduction to *E.E.* vol. I, no. 164. The English translation is attributed to Tyndale by J. F. Mozley, 'The English Enchiridion of Erasmus, 1533', *Review of English Studies*, vol. XX (1944), pp. 97–107. The quotations in my study are taken from this translation which was printed by Wynkyn de Worde. Pusino believes that Erasmus was influenced by Pico della Mirandola's writings (*Zeitschrift für Kirchengeschichte*, vol. XLVI (1928), pp. 75–96) in his *Enchiridion*; Hyma, 'Erasmus and the Oxford Reformers', *Nederlandsch Archief, loc. cit.* pp. 128–9 mentions Vitrier as a possible influence. Erasmus's religious point of view has been frequently attacked. See, for example, A. Hyma, *op. cit.* where some literature on this question is mentioned.

THE PHILOLOGICAL VIEW: ERASMUS OF ROTTERDAM

Continuing, he speaks of theologians who had never read Scripture but only Scotus. His advice is to study the theologians of old who are more profitable to the soul and more learned, who write a better style and know more of the mysteries of religion than modern theologians:

yf thou seke rather to haue thy soule made fatte than thy wyt to be vaynly delyted: study and rede ouer chiefly the old doctours and expositours whose godlynes and holy lyfe is more proued and knowen whose religion to god is more to be pondered and loked vpon whose lerning is more plenteous and sage also whose style is neyther bare ne rude and interpretacion more agreable to the holy mysteryes.[1]

As in the letter quoted above, where prayers are held to be useful as an introduction to theology and to style, so here the early Fathers will prepare the reader in these two disciplines. But whilst Colet condemned the reading of pagan literature, Erasmus cannot speak against the study of pagan poets and philosophers. Yet in his *Enchiridion* he wishes to limit the time devoted to these studies and repeats words which he had used as a warning against scholastic philosophy in his *Ratio Studii*: one should not grow old there, at the rocks of the Sirens, as it were. But he points out:

Those scyences fascyon and quycken a chyldes wytte and maketh hym apte aforehande meruaylously to the vnderstandyng of holy scripture.[2]

Thus the study of letters is preparatory to theology. Yet even profane *literae* should be understood in the same way as Holy Writ, namely as allegories.[3] Allegorical interpretation is made an instrument of literary interpretation. Erasmus makes no distinction between the method of study of profane and of sacred literature. There is no difference, Erasmus asserts, between Livy's History

---

[1] *L.B.* vol. v, col. 8 c-f, Wynkyn de Worde, 1533 fol. B vij^v and B viij^r.
[2] *L.B.* vol. v, col. 7 d-e, Wynkyn de Worde, fol. Bv^v. Cf. *Methodus, N.T.* 2, p. 16. A. Renaudet, *Études Érasmiennes* (1521–1529), Paris, 1939, p. 135, has drawn attention to this similarity between the *Enchiridion* and *Ratio*.
[3] *L.B.* vol. v, col. 7 f.

and the Books of Kings or the Book of Judges unless the allegorical meaning is fully understood:

> What difference is there whether thou rede the boke of kynges or of the iudges in the olde testament or els the history of Titus Liuyus so thou haue respecte to the allegorye in nere nother? For in the one that is to say Titus Liuyus be many thynges whiche wolde amende the comen maners: in the other be some thinges ye vngoodly as they seme at ye first lokyng on whiche also if they be vnderstande superstycially shulde hurte good maners.[1]

This praise of allegory, which has its forerunners in earlier humanists, caused Erasmus to consider Origen as the best expositor of the Bible, a view which he maintained throughout his life.[2]

The importance of this period for Erasmus is very great indeed. The observer can see the outline of things to come, although it may be doubtful whether Erasmus could from these rudimentary thoughts see the way which he would follow. The student can observe how Erasmus tries from every point of view to find a bridge which would connect *bonae literae*, humanism, and *sacrae literae*, theology. He attempts to make a case for the humanistic studies of the profane arts as the basis for the higher study of the mysteries of Holy Writ which are 'the principal part of theology'. He later retains this distinction but he is occupied only with the humanistic part and for this reason he could in later life maintain that he had never claimed to be a theologian.[3] Yet even for theology humanism is, in his opinion, of importance, for mystery must be 'fortyfyed with strengthe of eloquence and tempred with certayne swetnesse of speakynge'.[4]

---

[1] *L.B.* vol. v, col. 29 D–F, Wynkyn de Worde, fol. H vij^v.

[2] This aspect has been neglected by all those who consider Erasmus as a man of enlightenment. On Allegory see esp. *Enchiridion*, ch. 8, Canon 5, *L.B.* vol. v, cols. 27 ff. On Origen see, for example, *E.E.* vol. I, no. 181 (38–41), of ⟨c. December⟩ 1504; *Methodus* in *N.T.1*, fol. bbb 4^v; *E.E.* vol. III, no. 844 (252–4), of 15 May 1518.

[3] *L.B.* vol. IX, cols. 66 B, 1101 D, 1215 C–D. Cf. R. Pfeiffer, 'Humanitas Erasmiana', *loc. cit.* p. 1 and n. 1.

[4] *L.B.* vol. v, col. 30 A, Wynkyn de Worde, fol. H viij^r.

THE PHILOLOGICAL VIEW: ERASMUS OF ROTTERDAM

In 1501 Erasmus started with his work on a theological subject, for he wrote four volumes on St Paul's Epistle to the Romans. Our knowledge of this derives from a short remark at the end of his *Enchiridion* where he mentions that he has 'Adorned the Temple of the Lord with foreign treasures'.[1] But three years later, in a letter to Colet with whom he had not been in communication for years, he admits that the main reason for not completing his work on St Paul was his incompetence in Greek. These volumes have not been preserved and it is futile to speculate about their contents and on the persons who may have influenced Erasmus.[2] But obviously Erasmus had understood the significance of Greek and Hebrew for the study of the Bible, for he also tried to learn Hebrew, but 'frightened by the strangeness of the idiom' and considering 'his age and the insufficiency of the human mind to master a multitude of subjects' he gave it up.[3]

In this letter to Colet he stresses the influence of Origen on him.[4] Repeating the simile he had used in 1500 he says:

But although I am meanwhile engaged in a perhaps humbler work, nevertheless while I pass my time in the gardens of the Greek authors, I gather, as I go on, much that will be also of use in sacred studies. For this one thing I know by experience, that we cannot be anything in any kind of literature without Greek. For it is one thing to guess, and another to judge; one thing to believe your own eyes, another thing to believe other people's.[5]

With these words of 1504 he repeats and makes clearer than before what he had expressed in 1501:[6] that Greek is the basis of all literature be it sacred or profane. The study of Greek secular authors will further the knowledge of the Bible. The distinction

---

[1] *E.E.* vol. I, no. 164 (36–55)=*L.B.* vol. V, col. 66A–C.
[2] A. Hyma, *Nederlandsch Archief, loc. cit.* p. 128, suggests that John Vitrier was 'partly responsible'. His evidence (*E.E.* vol. I, nos. 163 (6) and 165 (9)) is hardly sufficient to warrant his conclusion.
[3] *E.E.* vol. I, no. 181 (34–8), of ⟨c. December⟩ 1504.
[4] *E.E.* vol. I, no. 181 (38–41).
[5] *E.E.* vol. I, no. 181 (88–93)=*N.* vol. I, p. 378. The last sentence is almost verbatim repeated in Erasmus's Preface to Valla's *Adnotationes* (*E.E.* vol. I, no. 182 (187–8), of ⟨c. March⟩ 1505).
[6] See above, pp. 127–8.

between the two disciplines literature and theology disappears, for Holy Writ can be considered as literature. Thus the method of their interpretation is identical. But it must be remembered that this theology deals with the literal meaning only, not with the divine mysteries, a part of theology which has its own method although it is derived from the results of the humbler branch.

It is, however, amazing to find that Erasmus repeats part of a sentence which he had written to Colet in 1499, namely that *bonae literae* 'ceased to be agreeable to him'. In 1504 he writes:

I beseech you [Colet] therefore to do what you can to help me in my craving for sacred studies, and to rescue me from that kind of *literae* which has ceased to be agreeable to me.[1]

A few lines after this sentence he mentions that a new edition of the *Adagia* is necessary because the first is full of misprints and poor in contents since at the time of its composition he did not know the Greek authors. The preparation of this new edition, it seems from the context, is not part of 'that kind of *literae* which has ceased to be agreeable' to him. He had, of course, stressed the importance of proverbs for the comprehension of Holy Writ. Yet again the reader is left in the dark about the exact meaning of *literae* which, like theology, may be of a twofold nature, the one desirable, and the other unnecessary, to Erasmus.

After he had written this letter to Colet he found in the Premonstratensian Abbey of Parc near Louvain a manuscript of Laurentius Valla's *in Latinam Novi Testamenti Interpretationem ex Collatione Graecorum Exemplarium Adnotationes*, a book that confirmed his view on the connexion between humanistic scholarship and theology. This work of Valla's, finished in 1449, had never been printed; Erasmus, advised by Christopher Fisher, who at that time was Papal Protonotary, published this manuscript on 13 April 1505.[2]

---

[1] *E.E.* vol. I, no. 181 (76–8)=*N.* vol. I, p. 377.

[2] Printed by Badius, Paris. Details on editions are given *B.Er.* 1, 2, p. 66; P. S. Allen, Introduction to *E.E.* vol. I, no. 182. Erasmus does not mention Valla in his letter to Colet. I therefore assume that he found the MS. after he had written the letter. Cf. *N.* vol. I, p. 378 n.

THE PHILOLOGICAL VIEW: ERASMUS OF ROTTERDAM

Valla[1] had given no outline of his method in his *Adnotationes*; this must be gleaned from the notes themselves. A few points may be mentioned here to illuminate his general views. Valla had recognized that in principle the New Testament, in the original Greek, is to be preferred to the Latin Vulgate. Therefore he had compared Greek manuscripts of the Bible (he mentions four manuscripts) with the Vulgate.[2] The results of this comparison are: the Latin text often differs from the Greek; there are omissions and additions in the Latin translation;[3] the Greek wording is generally better than the Latin one.[4] Valla demands from every translator the compliance with certain rules which are in his opinion not heeded in the Vulgate. These rules are: different words in Greek should be rendered with different words in Latin, while the same Latin word should be used for any one Greek word;[5] the construction of Greek sentences should be imitated in Latin as much as possible but the rules of Latin grammar and syntax must be observed;[6] guidance in respect of all linguistic questions is to be found in the works of pagan authors—as they lived earlier than the Christian writers their authority in matters of language is weighty.

There are people [Valla continues] who believe that theology is not subservient to the rules of grammar. But I say that theology must observe the usage of the spoken and especially of the written language. For what is more stupid than to corrupt the language used and to make it faulty? For when doing so one cannot be understood by one's fellow speakers. Nobody understands him who does not observe the property of the language.[7]

---

[1] Literature on Valla and Erasmus: see P. Mestwerdt, 'Die Anfaenge des Erasmus, Humanismus und "Devotio Moderna"', *Studien zur Kultur und Geschichte der Reformation*, vol. II (Leipzig, 1917), pp. 46ff. R. Pfeiffer, 'Humanitas Erasmiana', *loc. cit.* pp. 11–14. A. Renaudet, 'La critique Érasmienne et l'humanisme français', *Bijdragen voor Vaderlandsche Geschiedenis en Oudheidkunde*, vol. VII, no. 7 (1936), pp. 228–30.

[2] Quotations refer to the edition of 1505. In general only one reference will be given for each example. The above refers to Matth. xxviii. 8, fol. XII^r (middle).

[3] Fol. xxxi^r to I Cor. vi. 20; fol. xii^r to Matth. xxviii. 8.

[4] Fol. xxxiii^r to I Cor. xv. 31–2.

[5] Fol. xxx^r to I Cor. iv. 3–4; fol. iiii^v (A vi^v) to Matth. vii. 24.

[6] Fol. viii^r (B iii^r) to Matth. xviii. 24; cf., for example, fol. xxxiii^r to I Cor. xv. 31.

[7] Fol. iiii^r (A v r) to Matth. iv. 10.

THE PHILOLOGICAL VIEW: ERASMUS OF ROTTERDAM

Valla's intention was to correct the translation of the Vulgate and in this way to attain a double goal: to correct mistakes, and to refine the language in accordance with the requirements of classical authors.

Valla had yet another purpose in his *Adnotationes*. He wished to show the futility of those who expound the New Testament without knowing Greek. 'I am amazed', he writes, 'that people utterly ignorant of Greek have dared to write commentaries on Paul who spoke Greek, especially as so many commentators, Greek as well as Latin, who knew Greek have done so.' This general principle he tries to prove by an acrimonious invective against Thomas Aquinas's *Commentaries on St Paul*. It is related that after he had finished his commentaries, Thomas was visited by St Paul and was told that he understood his writings better than anybody else. Valla telling this continues:

was Paul better understood by Thomas than by Basil, Gregory of Nazianzus, Chrysostom? (Why do I mention Greek writers?) or by Ambrose, Jerome, Augustine? May I die if this is not a lie. For why did Paul not remind him of all his mistakes, especially of his ignorance in Greek?[1]

The views worked out slowly by Erasmus for years are closely set out in Valla's book. Thus the discovery of this manuscript is important for Erasmus, and I would call it the last stage of Erasmus's preparation for his work on the Bible. In the Preface[2] Erasmus defends Valla whose works he had appreciated even when in the monastery of Steyn.[3] He again refers to the Council of Vienne. But the main importance of this preface lies in the discussion of the question of principle, whether the grammarian or only the theologian is allowed to criticize the Latin translation of the Bible. Erasmus refers to Nicolaus of Lyra's criticism of St Jerome's version, a criticism based on the comparison of the Hebrew text with the Vulgate. The plea that Lyra, the theologian,

---
[1] Fol. xxxi^v to 1 Cor. ix. 13. Valla's scorn on Remigius is fol. xxxviii^v to 1 Thess. i. 8.
[2] Reprinted *E.E.* vol. I, no. 182; it is dated ⟨c. March⟩ 1505.
[3] Cf. above, p. 96.

is allowed to censure Jerome, but Valla, the grammarian, is not is according to Erasmus futile. Valla's position is, in Erasmus's view, stronger than Lyra's, for the Hebrew texts may, at Lyra's time, have been corrupted on purpose, while Valla had collated old and emended Greek manuscripts and written notes on the New Testament. These notes deal with discrepancies between the Greek and the Latin texts, or with inaccuracies caused by the translator, or with a lack of significance in the Latin version, or lastly with corruptions of the Latin wording.[1] Is this work, Erasmus continues, to be called theology or philology?

When Lyra discusses a form of expression, is he acting as a theologian or is he not rather acting as a grammarian? Indeed all this translating of Scripture belongs to the grammarian's part; and it is not absurd to suppose Jethro to be in some things wiser than Moses. Neither do I think that Theology herself, the Queen of all Sciences, will hold it beneath her dignity to be attended and waited upon by her handmaid, Grammar; which if it be inferior in rank to other sciences, certainly performs a duty which is as necessary as that of any.[2]

Erasmus thus clarifies the position. Theology and grammar have their own field. It is the task of the non-theologian to provide the translation of the Bible for the theologian. Grammar is therefore a discipline in its own right, humbler than 'Theology, the Queen of all Sciences', but a necessary prerequisite, a 'handmaid' only, but one not to be despised, for Erasmus's allusion to Jethro and Moses implies that the handmaid may be wiser than the mistress. It may, however, be asserted, Erasmus continues, that translation of Scripture does not depend on rules of grammar but on the inspiration of the Holy Spirit.[3] He expects that an attack against his philology will be based on the inspirational principle of Bible translation. Knowing the controversy between St Jerome and St Augustine, he asserts the right of the philologist, using

---

[1] *E.E.* vol. I, no. 182 (116–25).
[2] *E.E.* vol. I, no. 182 (128–35) = N. vol. I, p. 382.
[3] *E.E.* vol. I, no. 182 (138–9): '...totum interpretandi negocium de sacri Spiritus afflatu pendere....'

Jerome's words, in which the fundamental difference between the two methods is most pointedly set out:

> It is one thing to be a prophet and another to be a translator; in one case the Spirit foretells future events, in the other sentences are understood and translated by erudition and command of language.[1]

Erasmus, moreover, pours scorn upon the devotees of the inspirational theory who create a new honour for the theologian: 'these are the only people who are privileged to speak incorrectly.'[2] If the inspirational method were right could Jerome lay down rules for translation? Since the translators of the Old Testament made some mistakes why should not those of the New Testament have erred? Does not Jerome in his emendation of the earlier translation of the Bible admit that he often fails to change the old version? These passages should be corrected and they especially are chosen by Valla for criticism. 'Shall we ascribe to the Holy Spirit the errors which we ourselves make?' Even if the original translators made no mistakes, their words became corrupt and had to be emended by St Jerome. His version has again become perverted and has to be corrected with all the 'caution and moderation which is due to all books and above all to the sacred volume'. In this way Erasmus answers the opponents of textual criticism, who say: it is not right to make any change in the Holy Scriptures in which even the apexes have some mysterious meaning. It is, according to Erasmus, the duty of the philologist to reconstitute the text in its original splendour. Depravations which have come about for reasons of ignorance must be amended. These corruptions are evident from the fact that copies of the Vulgate vary greatly in their wording. Valla's work can in no way be compared with St Jerome's, who substituted a new version for an old one; Valla, however, 'collects his observations in a private commentary, and does not require you to change anything in your book, although the very variety we find in our copies is sufficient evidence that

---

[1] E.E. vol. I, no. 182 (141–3) = N. vol. I, p. 382 (I have changed Nichols's 'interpreter' to 'translator'); St Jerome, *Praefatio in Pentateuchum* (P.L. vol. XXVIII, col. 151A).

[2] E.E. vol. I, no. 182 (139–40) = N. vol. I, p. 382.

they are not free from errors. And as the fidelity of the old books is to be tested by the Hebrew rolls, so the truth of the new books requires to be measured by the Greek text, according to the authority of Augustine, whose words are cited in the Decreta.'[1]

This preface, elaborate in its detail and very elegant in its language, is a defence rather than an attack. Erasmus expected strong censure from the side of theologians who would come forward in protection of the official wording of the Vulgate and who at the same time would be opposed to the temerity of the humanist who dared to treat theological subjects. Erasmus wished to forestall these attacks. He understood perfectly that this could only be done if a theologian himself were made to hold his opinions. Therefore he quoted St Jerome and the Decreta. He thus made his position as unassailable as possible.

A year later Reuchlin in his *De Rudimentis Hebraicis* used his knowledge of Hebrew for a criticism of the Latin version of the Old Testament. Like Erasmus, he wished to secure his position. Though his preface is less elaborate and his style is clumsy in comparison with that of Erasmus, the spirit in which these prefaces are written is the same. It cannot be denied that Reuchlin may have known Erasmus's edition of Valla. Yet the similarities of these two dedicatory letters may have sprung from the fact that the same purpose guided both men. They expected to be subjected to immediate and bitter attacks. But the storm did not break. Neither the *Rudimenta Hebraica* nor the edition of Valla's *Adnotationes* were made the starting-point of the war against humanism.

There is, however, one great difference between Reuchlin's preface and that by Erasmus. Reuchlin emphasized the necessity of knowing Hebrew, the original language of the Old Testament. Erasmus, in the same vein, impressed the reader with his belief that one must know Greek for the New Testament. But at the

---

[1] *E.E.* vol. I, no. 182 (137–71) = N. vol. I, pp. 382–4. The sentence referred to is *Decreta*, dist. ix, c. vi, where it is erroneously ascribed to Augustine instead of to Jerome, *Praefatio in Pentateuchum* (*P.L.* vol. XXVIII, col. 152).

end, Erasmus, winding up all his arguments, not only stressed the importance of Greek for the understanding of Holy Writ, but added that a knowledge of literature was also essential. His words are:

They have neither sense nor shame who presume to write upon the sacred books, or indeed upon any of the books of the ancients, without being tolerably furnished in both literatures, for it may well happen, that while they take the greatest pains to display their learning, they become a laughing stock to those who have any skill in languages, and all their turmoil is reduced to nothing by the production of a Greek work.[1]

In these words all those thoughts are gathered together which could be detected in Erasmus's earlier writings, but most of them are also found in Valla's *Adnotationes*: the basis of comprehension of any literary work, sacred or profane, is the knowledge of Greek. The writer on either literature must know the other.

We can now see the development of Erasmus's thought much better than before. In 1500, in his Preface to the *Adagia*, he had recommended a knowledge of proverbs as useful for the two literatures, but Greek was still in the background. In 1504, before he had found Valla's *Adnotationes*, he had written to Colet: 'I pass my time in the gardens of the Greek authors, I gather, as I go on, much that will also be of use in sacred studies.'[2] We remember Valla's assertion that Christian authors must have followed the usage of heathen authors.[3] Thus the synthesis of sacred and profane literature, or of theology and humanism, is the only means of coming to a true understanding of both literatures. It is the aim of Erasmus's life to achieve this synthesis, which bridges the gulf between religious and worldly tendencies. That promotion of both profane and sacred studies which raises him above his fellow-humanists is praised by a close friend of Erasmus whose judgement carries, in my opinion, more weight than that of anybody who

---

[1] E.E. vol. I, no. 182 (193–8) = N. vol. I, p. 384.
[2] E.E. vol. I, no. 181 (88–90) = N. vol. I, p. 378 (quoted above).
[3] See above, p. 133.

THE PHILOLOGICAL VIEW: ERASMUS OF ROTTERDAM

relies on written documents only, for this witness knew by personal experience all the tendencies of his time. He is Thomas More, who in a letter to one of Erasmus's opponents, Edward Lee, writes:

I admit that I love Erasmus ardently for almost no other reason than that for which the whole Christian world likes him; for as a result of the indefatigable labours of him alone, more than through the erudition of almost anybody else in the last centuries, have students of *bonae literae* everywhere promoted profane as well as sacred studies.[1]

A similar statement is made by Mutian in 1520 and by Urban Regius in 1522.[2]

In 1505, after the edition of Valla's critical notes of the Bible, Erasmus went to England again. And it was here that he made a translation of the New Testament. John Colet, who in the meantime had become Dean of St Paul's,[3] lent him two Latin manuscripts which Erasmus could decipher only with difficulty: but nothing is known about the Greek manuscripts used. The loan of these manuscripts proves that Colet had no objection to the translation being made. It is even possible to conclude that he approved of it, for an amanuensis of his, Peter Meghen of Brabant, an excellent calligrapher, copied the translation twice; one of these copies, intended as a present for Colet's father, Sir Henry Colet, is in three volumes written on vellum in 1506 and 1509. John Colet's interest in Erasmus's translation is thus certain without a shadow of a doubt. In all probability Erasmus had convinced him of discrepancies between the Greek and Latin texts, though not of the importance of learning Greek for the interpretation of the Bible.[4]

---

[1] *Mo.E.* no. 75 (105–10), of 1 May 1519.

[2] *Mu.E.* no. 590 (p. 260), of 24 May 1520; *E.E.* vol. V, no. 1253 (15 ff.), of 4 January 1522.

[3] A discussion of the date is found in *E.E.* vol. I, no. 181 n. 18.

[4] For all details on the Colet MSS. see P. S. Allen, introductions to *E.E.* vol. II, nos. 373 and 384, esp. p. 182; cf. *idem*, *The Age of Erasmus* (Oxford, 1914), pp. 141–2. The MSS. are: (1) A copy written by Meghen in three volumes, the first containing all the Epistles dated

THE PHILOLOGICAL VIEW: ERASMUS OF ROTTERDAM

This translation is remarkable for the freedom with which changes in the wording of the Vulgate are made. Two Prefaces, those to the Epistles of St Peter and St John, clearly reveal Erasmus's method. Both are short, containing remarks about style. Peter's epistolary style is characterized: 'it is worthy of the prince among the apostles, full of apostolic authority and majesty, economical in words but filled with meaning.'

The argument relating to the First Epistle of St John is of greater importance since Erasmus had to discuss its authenticity. He does this only from the stylistic angle. The character of the language proves, in his view, that the epistle was written by the apostle: the same words are used in the Gospel as in the letter. Moreover, a criterion characteristic of John's style is stated, namely the repetition of a word in two or more consecutive sentences, such as the reiteration of the word 'world' in the following lines: 'Love not the world, neither the things that are in the world. If any man love the world, the love of the Father is not in him. For all that is in the world....'[1] Erasmus ends his argument with these words: 'Lastly there is in the whole of his language something that is so to speak more loquacious than in the language of the other apostles.'[2] It is noteworthy that Erasmus does not use theological reasoning to prove his point. He considers linguistic and stylistic peculiarities which may be used in profane literature in the same way. As to the Biblical epistle, it is the grammarian who advises the theologian; it is necessary to see these words within the perspective of Erasmus's work to comprehend their full significance. In the preface to a reissue of a

---

[1] November 1506, the second comprising St Luke and St John, dated 7 September 1509 (both in the British Museum), and the third containing St Matthew and St Mark, dated 8 May 1509 (in the University Library of Cambridge). A description of these MSS. is found in *British Museum, Catalogue of Western Manuscripts in the Old Royal and King's Collections* (1921), pp. 19–20. (2) A complete translation (containing Acts and Revelation as well) is in the Library of Corpus Christi College, Oxford.

[1] 1 John ii. 15–16.

[2] The texts are in the MSS. of the British Museum on (fol. pp. 6ʳ) and (ss 7ᵛ). These Prefaces were printed by Th. Martens, Louvain, November 1518 and in the second edition of Erasmus's *New Testament* of 1519 (pp. 501, 516), the text of which fully agrees with the translation of 1505–6. Cf. P. S. Allen, Introduction to *E.E.* vol. II, no. 384.

THE PHILOLOGICAL VIEW: ERASMUS OF ROTTERDAM

translation of Euripides's *Iphigenia* he compares this play with Euripides's *Hecuba* and, observing a difference of style in the two dramas, he raises the question whether the *Iphigenia* were written by Euripides, for there seems to be, he says in 1507, 'a change in the taste of the language and in the character of the poetry'.

For if I am not mistaken, it [*Iphigenia*] has a little more brilliancy and the diction is more free. In this respect it might seem like Sophocles; but again in the closeness of the arguments and in a sort of declamatory power of persuading and dissuading, it rather recalls Euripides as its parent.[1]

Thus it can be seen that the same method is employed for the discussion of sacred and profane literature. The synthesis of theology and humanism, as proclaimed by Erasmus, is derived from a method which is applied to both these disciplines. This method is humanistic; it is in many respects the dawn of modern scholarship.[2]

It is time to pause here and to review Erasmus's achievements. Without any doubt his work has grown in scope. It is universal, comprising the secular and sacred knowledge of his time. It is far wider than mere scholarship. In its demands it requires a new approach to life based on humanistic training. It is no good for the theologian, Erasmus asserts, to be trained in the sophisms of the schoolmen which make men contentious but not wise. A new and different education is needed to bring about a changed attitude towards the Bible. The new civilization in which there is no contrast between *bonae literae* and *sacrae literae* must have the Bible as a centre in which all wisdom is contained; but advancing towards this centre man has to study secular knowledge, without which he cannot arrive at any understanding of the divine word.

What is it that gives Erasmus this confidence that his conception of biblical studies is right? How is it possible for him to discover the truth of Holy Writ that has remained hidden during so many centuries? The answer is simple. He does not follow his own

---

[1] Dedicatory Epistle (*E.E.* vol. I, no. 208 (1–8)=*N.* vol. I, p. 431; cf. *E.E.* vol. I, no. 188). For Erasmus's secular translations see W. Schwarz, 'The Theory of Translation in Sixteenth-Century Germany', *Modern Language Review*, vol. XL, pp. 291 ff.

[2] For similar examples in Erasmus *N.T.1* see below, p. 153.

THE PHILOLOGICAL VIEW: ERASMUS OF ROTTERDAM

individual reasoning, which, in its subjectivity, can prove nothing. He has found a new approach which is based on a philological method. This is most clearly pointed out in the prefaces to his edition of the New Testament, which more than any other work of its time endangered the authority of the Vulgate, the official version of the Bible.

Erasmus's New Testament was first printed by Johannes Froben at Basel in February 1516; but Erasmus was dissatisfied with this edition and republished it, completely revised. In the first edition the text of the Bible is preceded by a dedicatory epistle to Pope Leo X, *Paraclesis*, *Methodus*, and *Apologia*. The second edition of 1519 reprints the dedicatory epistle, *Paraclesis* and *Apologia*, but adds a letter from Leo X to Erasmus recommending this Bible. Furthermore, *Methodus*, which occupies a little over eight pages in the first edition, has been enlarged to almost fifty pages with the title *Ratio seu Compendium Verae Theologiae*; a treatise called *Capita Argumentorum contra Morosos quosdam et Indoctos* is added, containing 111 points in justification of this method.[1] The Greek and Latin texts of the Bible are printed side by side, followed by a Preface to the *Adnotationes* and finally by the *Adnotationes*. While in the first edition Erasmus keeps the wording of the Vulgate with 'some superstitious fear'[2] he is much freer in the second edition, where he has his own version of 1505/6 printed.[3]

---

[1] Editions of these Prefaces: Letter to Pope Leo X=*E.E.* vol. II, no. 384. Letter from Pope Leo X=*E.E.* vol. III, no. 864. Bibliography on *Ratio*: P. S. Allen, Introductions to *E.E.* vol. III, no. 745, and vol. V, no. 1365; *B.Er.* I, pp. 167–9; *B.Er.* 2.

[2] *Adnotationes* to John i. 1, p. 353.

[3] For editions see *B.Er. I*, part 2, pp. 57–66. For details concerning Erasmus's use of MSS., his work for the editions, his supervision of the printers, etc. see: P. S. Allen, Introductions to *E.E.* vol. II, nos. 373, 384; *E.E.* vol. III, no. 864; cf. P. S. Allen, *Erasmus*, pp. 44–9, 68–70; R. Wackernagel, *Geschichte der Stadt Basel*, vol. 3 (Basel, 1924), pp. 132 ff., esp. pp. 153–8, 222–3, 230 ff., 421–6. In this book by Wackernagel valuable contributions are found on sixteenth-century book production, book trade, relationship between printer and scholar in general and between Erasmus and Froben in particular, Erasmus's editions etc. Descriptions of, and discussions on, the New Testament: A. Bludau, 'Die beiden ersten Erasmus-Ausgaben des neuen Testaments und ihre Gegner', *Biblische Studien*, hrsggb. O. Bardenhewer (Freiburg Br., vii, 1902), pp. 12 ff.; A. Renaudet, *Études Érasmiennes, 1521–1529* (Paris, 1939), pp. 138 ff., 153 ff.

THE PHILOLOGICAL VIEW: ERASMUS OF ROTTERDAM

It is not the purpose of this study to discuss Erasmus's Bible in detail or to single out special parts for praise or blame. Only the main principles of the work, which derive logically from Erasmus's thought in general, shall be shortly described.

Even the arrangement of the Biblical texts shows the intention of the author: the Greek text is printed side by side with the new translation; notes at the end of the work explain why the new version differs from the text of the Vulgate. Thus the reader is offered a new text, and only if he looks up the Annotations at the end can he find the traditional wording—criticized and rejected for certain reasons. One may admit, Erasmus argues, that the Greek manuscripts are as corrupt[1] as the Latin ones. For more than a thousand years, he says in the *Apologia*, there have been differences between the Greek and the Latin texts due to the ignorance, negligence and thoughtlessness of scribes. But errors in the tradition of one manuscript can often be corrected from another. A comparison between Greek and Latin manuscripts can often lead to the restoration of the original text. For this reason, Erasmus continues, Augustine and Jerome suggested that the Greek wording should be consulted.[2] Erasmus arrives at the conclusion that the readings of the Vulgate may in some places be better than those of the Greek manuscripts. For this an example may be given: Only the Greek manuscripts have the addition to the Lord's Prayer (Matth. vi. 13) 'for Thine is the Kingdom, and the Power, and the Glory, for ever, Amen'. Erasmus points out that his translation of the New Testament follows the Greek text and that he therefore has rendered these words, although in his opinion they were added by the Greeks.[3] But he makes his rejection of these words clear in the Annotations. The insertion of this phrase which does not belong to the text shows that he wished to avoid competition with the Vulgate, for his new translation,

---

[1] Cf. E.E. vol. III, no. 843 (363–70), of 7 May 1518.
[2] *Apologia*, N.T.2, pp. 64–5. *Contra Morosos*, N.T.2, nos. 69–71, p. 76.
[3] *Apologia*, N.T.1, fo. bbb 7ᵛ = N.T.2, p. 66; cf. E.E. vol. III, no. 860 (58–61), of 26 August 1518 and Erasmus's notes to Matth. vi. 13 in his editions of the New Testament.

though based on the original, is not faultless, since the Greek manuscripts may have been changed. It is worth noting that these words which, in Erasmus's view, are added to the Lord's Prayer, are printed in smaller type than the ordinary text from the second edition onwards. The attention of the reader is therefore directed at once towards these words.

Erasmus's reason for rejecting the genuineness of the Greek reading may serve as an introduction to his method. It is that the words[1] 'for Thine is the Kingdom...' though found in all Greek manuscripts are not found in the Latin ones.[2] Valla, Erasmus writes, believed that the Latin manuscripts have a curtailed text,[3] he disagrees with Valla on this point—if it were true, one must necessarily admit that all the Latin writers and the whole Roman Church have not known an appreciable part of this prayer. Moreover, Valla's assumption requires an explanation of the fact that Jerome left these words out without even mentioning it in his commentary.

From this example we can draw the following conclusions concerning Erasmus's method of reconstituting the text:

(1) Erasmus compares several Greek manuscripts, indeed as many as he can lay his hands on to discover variants in their readings. (It is well known that he looked for manuscripts everywhere during his travels and that he borrowed them from everyone he could.)[4]

(2) Erasmus attempts to verify the reading of the Vulgate by means of old Latin manuscripts, as, for example, those lent to him by Colet in 1505. This work, together with the comparison

---

[1] I generally follow the *Adnotationes* of the second edition of 1519.

[2] In the first edition Erasmus makes a bad mistake. He introduces a 'recent' writer Vulgarius who is also mentioned in the title-page. This is a misunderstanding for the Greek for 'Bulgarian' as found in connexion with Theophylact, the Bulgarian Archbishop (see R. B. Drummond, *Erasmus, His Life and Character*, vol. I (London, 1873), p. 315). In the second edition of the New Testament, 'Vulgarius' does not occur on the title-page but it is still found in the notes, e.g. to Matth. vi. 13 and Matth. i. 19. In the third edition he seems to be finally eliminated.

[3] *Adnotationes*, fo. iiii^r = A vi^r.

[4] For details and literature see P. S. Allen, Introduction to *E.E.* vol. II, no. 373. Cf., for example, *Contra Morosos*, no. 42, *N.T.*2, p. 73; *Apologia*, *N.T.*2, p. 64.

THE PHILOLOGICAL VIEW: ERASMUS OF ROTTERDAM

of the Vulgate with the Greek text, furthers his wish to reconstitute the text that has become corrupt through the ignorance of copyists. Attacked for his change of the Lord's Prayer he writes in 1519:

We change in the language of the apostles not even a syllable, indeed we reconstitute their language corrupted through the fault of copyists or otherwise. Nobody should cry: such a man corrects the Gospel, such a man emends the Lord's Prayer, but he should say, he purges the manuscripts of the Gospels from mistakes.[1]

(3) Erasmus uses the Fathers of the Church as independent witnesses for the early text of the Vulgate. In the dedicatory letter to the pope he mentions that the special care due to the sacred writings caused him not only to compare 'the oldest and most correct manuscripts' but also to 'run through all the writings of the old theologians and to trace from their quotations and expositions what each one of them had read and changed'. Needless to say, he takes corruptions in the manuscripts of the Fathers into account.[2] It is important to realize that Erasmus was the first to discover this very important principle for textual criticism, one that has been used ever since.[3] In the above example he proves the correctness of the Latin text through a reference to the early Fathers whose wordings agree with the Latin text and not with the Greek. Often, however, there is agreement between the Greek manuscripts and the commentaries of the Fathers. This, in Erasmus's opinion, proves that the 'schismatic' Greeks have not purposely corrupted the Greek text of the Bible.[4] Thus the Greek text may be used for the emendation of the Latin version.

[1] *Contra Morosos*, no. 20; cf. nos. 52, 53. *N.T.2*, pp. 71, 74.

[2] The dedicatory letter is reprinted in *E.E.* vol. II, no. 384 (49–59), of 1 February 1516. For a corruption in Augustine's text see, for example, note to Matth. xi. 30, *N.T.2*, p. 43; *N.T.3*, p. 46 (referring to *P.L.* vol. XL, col. 124).

[3] E. Nestle, *Introduction to the Textual Criticism of the Greek New Testament*, translated by W. Edie (London, 1901), p. 146, speaks of Francis Lucas of Bruges who in his notes of 1580 used this method 'for the express purposes of textual criticism'. But Erasmus was obviously the discoverer of this principle which seems to have been used by Cardinal Sirlet in about 1551–5. See H. Höpfl, 'Kardinal Wilhelm Sirlets Annotationen zum Neuen Testament', *Biblische Studien*, vol. XIII (Freiburg, Br. 1908), pp. 51–3.

[4] Cf. *E.E.* vol. III, no. 843 (342–62), of 7 May 1518.

THE PHILOLOGICAL VIEW: ERASMUS OF ROTTERDAM

(4) Erasmus attempts to explain the reason why the addition to the Greek text was made. He believes it was done for the sake of the solemnity of the prayer as the Latins often add 'Gloria patri' to the Psalms.[1] This argument again reveals the humanist who critically compares the texts and even believes in the possibility of discovering the reason for changes. This method is, of course, not limited to Holy Scripture, it is the philological method which Erasmus employed for the interpretation of every literary work.

Erasmus justifies his own translation by quotations from the Fathers. He can thus claim that his version is based on the authority of the Fathers. As an example of this I would like to mention his translation of Greek 'logos' with 'sermo' instead of the common translation 'verbum'.[2] Christ, Erasmus maintains, is called 'sermo' by the following Fathers: Cyprian, Augustine, Hilary, Ambrose, Lactantius, Claudian, Prudentius (in his hymns), and by the more recent theologians Thomas Aquinas, Cardinal Hugo, Bede, Remigius, Anselm, and the ordinary gloss. His text 'sermo' is therefore well founded within the tradition of the Church. His change of the wording of the Vulgate is not an act of heresy. He therefore can, from his point of view, say:

first of all we testify (and we wish it testified everywhere) that we do not intend to depart a finger's breadth from the judgement of the Church.[3]

This, however, does not mean that he is prepared to consent to the tradition and customs of the Church as they were at his own time. There is one authority only, which is beyond human criticism—the Bible, especially the New Testament which alone contains 'eternal wisdom' and from which alone true theology is derived.[4]

---

[1] *Apologia*, N.T.1, fol. bbb 7ᵛ = N.T.2, p. 66.
[2] This change was not yet made in the first edition.
[3] *Contra Morosos*, no. 1, N.T.2, p. 69.
[4] *Paraclesis*, N.T.1, fol. aaa4ʳ = N.T.2, p. 7. Cf. *Ratio*, N.T.2, p. 59. A discussion on 'philosophie du Christ et réforme religieuse' is found in, for example, A. Renaudet, *Érasme, Sa pensée religieuse et son action d'après sa correspondance, 1518–1521* (Paris, 1926), pp. 1ff.; J. B. Pineau, *Érasme, sa pensée religieuse*, pp. 50ff.

THE PHILOLOGICAL VIEW: ERASMUS OF ROTTERDAM

This is the fundamental belief of Erasmus, which resolves itself into the question of how to interpret the Divine Books in order to discover the truth contained in them. It is obvious that a quest for the truth of Holy Writ must first of all begin with a review of earlier commentators. There was, as mentioned above, a tradition of interpretation claiming an unbroken succession from Christ until the days of Erasmus. The commentaries in this tradition were within the framework of the Church and any deviation from it meant breaking away from the body of the Church. We have seen that Erasmus did not want to abandon the early tradition of the Fathers of the Church. He claimed the right to examine whether 'the celestial philosophy of Christ had not been vitiated' by human thought.[1] As a practical example, he mentions canon law which, though it takes its origin from the purest spring, the Bible, may in part not correspond with Christ's divine wisdom but rather with human frailty, which is dependent on conditions at the time when the individual law was drafted.[2] His note to Matthew xi. 30: 'For my yoke is easy, and my burden is light', may serve as an example of Erasmus's method.[3] From the comparison of the words of the Vulgate with the Greek text he moves on to mention the views of the Fathers, especially of St Jerome who had said that it is easier to bear the yoke of the evangelical law than that of the Mosaic law. The only burden which Christ laid on man is, Erasmus continues, charity. Christ's philosophy is in accordance with man's nature[4] and can therefore easily be borne. But exactly as the Jews aggravated the burden—here Erasmus changes the construction of his sentence by inserting one word which destroys the exact parallelism since it changes the confirmation expected in the second half to a warning, namely: so Christians should *beware* lest human orders and additions to

---

[1] *Ratio, N.T.2*, pp. 28, 56. A discussion on Erasmus's *Ratio* is found in A. Renaudet, *Études Érasmiennes, 1521–1529* (Paris, 1939), pp. 139 ff.

[2] *Ratio, N.T.2*, pp. 27–8.

[3] This note is not yet in *N.T.1*. I follow the third edition of 1522 which is an enlargement of the second edition.

[4] Cf. *Paraclesis, N.T.1*. fol. aaa 5ᵛ = *N.T.2*, p. 10.

dogma may make the burden heavy.[1] Erasmus speaks of the strong tendency of human nature which is bound to enlarge any burden originally added with pious intentions. He mentions 'custom whose tyranny is very great' and he points out that the Church, 'cut in pieces and tormented by disagreements with heretics', added further burdens. Thus the burden of the 'evangelical law' has increased beyond limit. But this development is going on, for, as Erasmus explains, at last it has come to pass that some opinions of the Schoolmen enjoy almost equal authority with the articles of apostolic faith. A sentence not understood is turned into a frigid syllogism, and so a human definition of little value ('constitutiuncula') is made an article of faith. 'The lifetime of a man is not long enough for these pseudoquestions and useless labyrinths of subtleties. When shall we find out what a Christian life is if even octogenarians have only learned to doubt?' As the result of this development the Christian people are 'ensnared by so many laws, so many ceremonies, so many traps, oppressed by the tyranny of princes, bishops, cardinals, popes, and much worse, by their satellites'. Erasmus enumerates the burdens imposed upon man, such as ceremonial dress, fastings, festivals, rites, matrimonial laws, and so on.

This argument, which also criticizes religious institutions, is based on the belief that everything has become depraved because man neglects to retain the perfect and changes it for the worse. Once a beginning has been made with this change, there is no end to it. A comparison with his early *Antibarbari* shows that fundamentally there is no change or break in Erasmus's development. Then, in the *Antibarbari*, he insisted that the decay of civilization is man's fault and can be corrected by man; now in his New Testament he applies this thought to every sphere of life: religious and secular powers are under his review. He is ready to criticize the tradition of the Church as far as its historical development is

[1] *Adnotationes, N.T. 3*, p. 47: 'Sed quemadmodum apud Iudaeos legem per se molestam aggravabant hominum constitutiones, ita cavendum est etiam atque etiam, ne Christi legem per se blandam ac levem, gravem et asperam reddant humanarum constitutionum, ac dogmatum accessiones.'

THE PHILOLOGICAL VIEW: ERASMUS OF ROTTERDAM

concerned, but he believes that Christianity (which means the Church for him) remains unimpaired. Christ's philosophy is in accordance with human nature: 'since human nature has fallen to ruin, Christ's philosophy deals with almost nothing else but with nature's reconstruction to innocence and sincerity.' The task for Erasmus is therefore to return to the holy truth which has been vitiated by man. His New Testament is an essential step in this direction: it restores the text of the Bible to its original wording as it was read by the early Fathers of the Church, who are for Erasmus the representatives of the old theology. This conception makes it imperative that all the intervening stages should be left out. The people of the Middle Ages could not, in their ignorance, see the purity of God's word. 'The brightness of eternal truth', he says, 'is reflected differently by a smooth pure mirror, differently by iron, differently by a most limpid spring, differently by a turbid pool.' It is the task of mankind to 'preserve that truly sacred anchor of evangelical doctrine to which we may have recourse in so great a darkness of human affairs'. The scope and goal of Erasmus's criticism of tradition is clear: it is to censure everything that has changed the original meaning of 'divine wisdom'.[1] As every human being errs, Erasmus thinks he is entitled to view critically everybody who has expounded the Bible: 'it is not insolent', he asserts, 'to dissent from the most acknowledged authors for proper reasons.'[2] And at another place: 'the Fathers of the Church were human beings, they were in ignorance of some things, they were dreaming and prating idly in others, and sometimes they were asleep.'[3]

There is no authority left which Erasmus does not criticize at one place or another in his annotations to the Bible. Yet he does not neglect the Fathers and does not arrogate to himself the science of Holy Scripture. Their commentaries further the work of the student, but they should not be read without criticism.[4]

---

[1] *Ratio, N.T.2*, p. 28.   [2] *Contra Morosos*, no. 61, p. 75.
[3] *Ratio, N.T.2*, p. 60; cf. E.E. vol. II, no. 456 (117–20), of ⟨22?⟩ August 1516.
[4] *Ratio, N.T.2*, pp. 59–60.

THE PHILOLOGICAL VIEW: ERASMUS OF ROTTERDAM

The highest authority is Christ's [he maintains]. He is the only Doctor created by God the Father; such an authority has been given to none of the theologians, to none of the bishops, to none of the popes or of the princes.[1]

Thus Erasmus has no scruples about attacking the Fathers of the Church and, of course, the sophisms of the Schoolmen in his annotations. He knew well that he exposed himself to attacks from every side through this criticism of the past. Every school of thought would have an axe to grind.

I know well [he writes] that some are annoyed at my dissension from weighty writers. Those who favour Augustine, gnash their teeth since I do not agree with him on every point. Addicts of Thomas are grieved at my attacks. The minorites would like to see neither Lyra nor Scotus refuted. But if we accept this proposition, we are not at liberty to dissent from anybody even though they dissent among themselves. In this work truth is to be respected more than authority.[2]

'Truth is to be respected more than authority', says Erasmus. 'Though I revere St Jerome as an angel and though I honour Lyra as a teacher, yet I worship truth as God', Reuchlin wrote a few years earlier. The same attitude moves both men. The quest for truth compels them to stop nowhere in their search and to examine critically the thoughts of even the greatest authorities of the Church. When editing Valla's *Adnotationes* in 1505 Erasmus had asked for freedom to criticize. In his New Testament he makes full use of this freedom regardless of attacks that might be levelled against him.

Erasmus's criticism of earlier commentators is fundamentally based on the belief that without knowledge of the three holy languages no one should attempt to interpret Holy Writ.[3] It is necessary to read the New Testament in Greek and to interpret the Greek text, and not the Latin Vulgate which gives rise to

---

[1] Annotations to Matth. xvii. 5 (ipsum audite) not yet in *N.T.1*.

[2] *Contra Morosos*, no. 60, p. 75: in hoc opere, veritatis maior quam autoritatis erat habenda ratio.

[3] Cf. *Apologia, N.T.1*, fol. bbb 6ʳ = *N.T.2*, p. 64.

many mistakes. Therefore the interpretation of the Schoolmen may be neglected as they did not know Greek. Matthew i. 19 may serve as an example for Erasmus's attitude; it shows Erasmus the philologist in his critical view on the Vulgate. The Authorized Version reads: 'Then Joseph her husband being a righteous man and not willing to make her a public example....' It is the last words 'make her a public example' (in Latin 'traducere') which elicit a note of more than one page from Erasmus. He expresses his amazement that the translator of the Bible who neglects polished language should yet have used the elegant word 'traducere' as a translation for Greek παραδειγματίσαι. Erasmus renders it with 'diffamare' in the first edition of the New Testament and with 'infamare' in the second edition. The reason for the change of the Vulgate text is the diversity of meanings of the Latin 'traducere', of which the most elegant and rarest sense is that used in the Vulgate. Erasmus adds quotations from Roman writers where this word has the same meaning. Besides, he adduces parallels from the New Testament where the same Greek word παραδειγματίζειν is used, shows from the context that his interpretation makes sense, and adds to this a quotation from Augustine as a testimony for his interpretation.[1] Yet Erasmus is not satisfied with this. He makes clear that the word 'traducere' has given rise to wrong comments from famous theologians, especially from Peter Lombard, a man 'upright and, according to his time, erudite'. His mistake, disgraceful though it is, must be imputed to the time rather than to its author from whose work unfortunately 'not swarms but oceans of never-ending questions have burst forth'. But how could the error be avoided, Erasmus writes, 'at a time when Greek and even Latin to a large extent was extinct, when Hebrew *literae* were more than dead, and when almost all the authors of antiquity were obliterated'?[2]

A similar judgement is found in the note to Romans i. 4 where Thomas Aquinas is praised as a man who is 'great not only for his century', who is sounder and more erudite than the later

---

[1] *P.L.* vol. XXXIII, col. 657.   [2] *N.T.*2, note to Matth. xi. 30.

theologians, and who made use of languages and *bonae literae*. The reason why great men like Thomas are in doubt about the meaning of St Paul is partly the obscurity of the words and sense, partly 'the misfortune of the times in which *bonae literae* had almost completely disappeared'.[1]

In these passages we again see a reminiscence of Valla's view that the decay of language is the reason for the decay of civilization. We recognize Erasmus's praise of humanistic scholarship implicit in his scathing remarks on the ignorance of earlier times. We note it is his knowledge of Greek that is used by Erasmus to overthrow the interpretations of the Schoolmen. Erasmus demands that the theologian should learn Greek instead of scholastic philosophy. His New Testament advocates a completely new course of studies which, through the training in the three holy languages and secular *bonae literae* as well as in the Fathers of the Church, leads to the interpretation of the Bible on humanistic principles.

The basis of such an interpretation is a knowledge of the meaning of the Greek or Hebrew words which gives a clue to the understanding of the passage.

> Those who narrate the sense [he says in the *New Testament* (Preface to the Annotations)] are often forced to unfold the meaning of the words; in the same way we are sometimes forced to lay open the full content of the sentences while we busy ourselves with the unfolding of the meaning of words.[2]

The interrelation of the meaning of the single word and that of the whole sentence, and the significance of the context for the meaning of the word are clearly seen by Erasmus. Different situations demand the use of different words. Each person's language has its own particular style and vocabulary; everybody speaks in a different way to people of different standing. Stylistic

---

[1] A similar praise is found in Erasmus's *Apologia*, *N.T.1*, fol. bbb 8ʳ=*N.T.2*, p. 67 where, however, Thomas is said to have known Latin only. Cf. *Ratio*, p. 16, *Apologia ad Iacobum Fabrum Stapulensem*, of 1517, L.B. vol. IX, col. 24 E–F. For Colet's censure of Thomas see *E.E.* vol. IV, no. 1211 (429ff.), of 13 June 1521.

[2] *N.T.1*, pp. 227–8=*N.T.2*, fol. aa 3ᵛ=*N.T.3*, fol. aa 3ᵛ.

THE PHILOLOGICAL VIEW: ERASMUS OF ROTTERDAM

observations of this kind were used by Erasmus for the characterization of the style of Greek writers whom he translated into Latin in January 1506.[1] In 1516 he remarks on the style of Biblical writing. He follows Jerome when he, for example, says that Paul's Greek is not elegant,[2] and that his style in the Epistles to the Ephesians and to the Hebrews differs from that in his other Epistles.[3] This philological method is used with consistency in the New Testament.

Something will be contributed to the understanding of Holy Scripture, [Erasmus says] if we carefully weigh not only what is said, but also by whom and to whom it is said, with what words it is said, at what times and on what occasion, what precedes and what follows. For, indeed, it is befitting for John the Baptist to speak differently from Christ. The ignorant crowd is enjoined in a different manner from the apostles.[4]

The individual situation must be clearly grasped for the understanding of the text of the Bible. For this the knowledge of history is essential, but also that of botany and zoology, and all the other auxiliaries which help towards an understanding of the events described. If the commentator is lacking in the knowledge of these secular sciences and is satisfied to consult dictionaries only, he will make of a tree a quadruped, of a gem a fish and many other ridiculous things of this kind.[5]

The exposition of the meaning of every word within its context leads to another important conclusion. There are obscure words in the New Testament which have been interpreted in the most abstruse way as if they referred, for example, to the *Sententiae* of Peter Lombard. The meaning of such words must of necessity correspond to the meaning of the context and in this way to the meaning of the whole New Testament.[6] Therefore the interpreta-

---

[1] See above, p. 141.
[2] Note to 1 Cor. iv. 3. This note is much enlarged in the second and third editions. It refers to St Jerome, *P.L.* vol. XXII, cols. 1029–30. Cf., for example, *E.E.* vol. III, no. 844 (82 ff.), to Eck, of 15 May 1518.
[3] First note to Ephes. and note to Hebr. xiii. 24. Cf. above, p. 140, where Erasmus's remarks on the style of St John's letter are discussed. For some details see A. Bludau, *loc. cit.* pp. 50–1.
[4] *Ratio, N.T.2*, p. 24; cf. *ibid.* p. 57.
[5] *Ratio, N.T.2*, pp. 17–18. [6] *Ratio, N.T.2*, p. 57.

tion of every sentence will be confirmed by parallels from other sentences of the New Testament. These parallels are thus given by Erasmus for philological reasons and not for the purpose of solving theological questions. Erasmus's task is to speak about words, about the smallest details, in which, because of their smallness, even the greatest theologians sometimes blunder. It is reminiscent of Reuchlin when Erasmus maintains that 'great mysteries of divine wisdom are contained in syllables, even in the little apexes of the letters'.[1] Important questions are those concerning the different usages of idiomatic and figurative phrases in different languages, tropes, allegories, parables, and all the rhetorical forms of speech. These are peculiarities of the language which one must know if mistakes are to be avoided. It is such small philological details which, more than anything else, open the way to an understanding of Holy Writ.

For this is the aim: to understand the truth once revealed to mankind. Erasmus does not wish for innovations. It is the restoration of the old that he desires. 'We restore the old and exclude novelty', he says in his New Testament.[2] He knows well that for the restoration of the old, a new translation of the text of the Latin Vulgate is required. The principles of this translation are clear, but a few words may be said about its technique and, later, about the relationship of this new text to the Vulgate.

The technique of Erasmus's version tallies with his humanistic tendencies. The language of the translation is free from ungrammatical constructions. If the language of the apostles is not elegant, there is no reason to imitate them in it.[3] Some critics of this view assert, however, that Holy Scripture should not be subjected to

[1] Preface to *Adnotationes, N.T.1*, pp. 226–7=*E.E.* vol. II, no. 373 (111–15). Cf. *Contra Morosos*, no. 14, p. 71, *E.E.* vol. V, no. 1309 (59–61), of ⟨*c.* December⟩ 1515.
[2] *Contra Morosos*, no. 65, p. 76; cf. *Apologia, N.T.1*, fol. bbb 6ᵛ=*N.T.2*, p. 64; *E.E.* vol. II, no. 456 (77–8), of ⟨22?⟩ August 1516 to Henry Bullock: 'Neque enim nos novam prodimus aeditionem, sed veterem pro virili restituimus...', and almost the same wording in *Apologia in Dialogum Jacobi Latomi* (L.B. vol. IX, col. 104, no. 107; *E.E.* vol. IV, no. 1153 (185–6), of 18 October 1520).
[3] *Contra Morosos*, no. 16, p. 71.

THE PHILOLOGICAL VIEW: ERASMUS OF ROTTERDAM

the rules of Donatus.[1] 'These people say: "God is not offended by solecisms".' But these perversions of language though tolerated by Augustine often cause ambiguities and even perversion of the meaning.

Yet God hates the haughty solecists who attack those who speak correctly and who themselves do not wish to learn a more correct diction and like a dog in the manger do not let the others do so who do.[2]

Augustine's authority cannot be used for the defence of solecisms. Augustine believed in the necessity of using ungrammatical constructions for the multitude which would otherwise not understand the speaker. Nowadays, however, the crowds do not speak Latin but their own native languages.[3] The translation must be free from every possible misunderstanding; therefore idiomatic and figurative expressions should not be rendered word for word.[4] Erasmus asserts that he has not changed the original for the sake of elegant diction but in order to make the rendering faithful and the sense clear.[5]

Language consists of two parts [he says], namely words and meaning which are like body and soul. If both of them can be rendered I do not object to word-for-word translation. If they cannot, it would be preposterous for a translator to keep the words and to deviate from the meaning.[6]

He is, of course, aware of the shortcomings of every translation. In every version one meaning only can be rendered, even if the original wording allows of more than one interpretation. Erasmus tries to remove this deficiency through his notes, in which he mentions all the possibilities but points out which one he prefers, leaving the final judgement to the reader.[7] It is the philological

---

[1] *Contra Morosos*, no. 11, p. 70. Cf., for example, *E.E.* vol. III, no. 843 (88 ff.), of 7 May 1518.
[2] *Contra Morosos*, no. 7, p. 70. Cf. *Apologia, N.T.1*, fol. bbb 8ᵛ = *N.T.2*, p. 68.
[3] *Contra Morosos*, nos. 8 ff., p. 70.
[4] *Ratio*, pp. 50 ff.
[5] *Apologia, N.T.1*, fol. bbb 7ʳ = *N.T.2*, p. 65.
[6] *Contra Morosos*, nos. 28–9, p. 72; cf. *Apologia, N.T.1*, fol. bbb 7ᵛ = *N.T.2*, p. 66.
[7] Preface to *Adnotationes, N.T.1*, p. 226 = *E.E.* vol. II, no. 373 (54–6); cf. *Apologia, N.T.1*, fol. bbb 7ᵛ = *N.T.2*, p. 66.

method which, as has been shown above, he uses in his notes as the basis of his interpretation. From this point of view Erasmus must be opposed to what, in the first chapter, has been called 'the inspirational method'. Those who believe in inspiration as being the ultimate source of the translation of the Bible must reject a new version based on philology. Erasmus makes some remarks against those who claim that the Septuagint is inspired.[1] In his New Testament he expresses his view: 'how was it possible that Jerome censured and changed the old Latin translation which was almost sacrosanct? How moreover could Cyprian, Ambrose and Augustine have different readings at their disposal?' Erasmus refers to the difference between a prophet and a translator as pointed out by St Jerome. But at the end of his brief remarks he speaks with irony against those who allow everything to a translator of old but nothing to a contemporary translator.

If they require inspiration of the Holy Spirit for the translator what is to prevent the presence in us of the Holy Spirit who is common to all Christians and who aids those more abundantly who have also added their own diligence?[2]

Erasmus obviously recognized that no reasoning is valid against those who advocate the inspirational method of translation. Since their fundamental attitude was so different from his own, no understanding with them was possible and his criticism was bound to result in one or other of the following reactions: hatred because of his wit and blasphemy, or contempt because of his inability to understand them.

But before we discuss the reaction to Erasmus's New Testament a passage may be quoted which sums up his 'plan of work' about which he writes in a letter of 26 August 1518:

Having first collated several copies made by Greek scribes, we followed that which appeared to be the most genuine; and having translated this into Latin, we placed our translation by the side of the Greek text, so

---

[1] *E.E.* vol. III, no. 843 (336–41, 478–90) of 7 May 1518.
[2] *Contra Morosos*, nos. 21–7, pp. 71–2.

that the reader might readily compare the two, the translation being so made, that it was our first study to preserve, as far as was permissible, the integrity of the Latin tongue without injury to the simplicity of the apostolic language.

Our next care was to provide, that any sentences, which had before given trouble to the reader, either by ambiguity or obscurity of language, or by faulty or unsuitable expressions, should be explained and made clear with as little deviation as possible from the words of the original, and none from the sense; as to which we do not depend upon any dreams of our own, but seek it out of the writings of Origen, Basil, Chrysostom, Cyril, Jerome, Cyprian, Ambrose, or Augustine. Some annotations were added (which have now been extended), wherein we inform the reader, upon whose authority this or that matter rests, relying always upon the judgement of the old authors. We do not tear up the Vulgate edition—which is however of uncertain authorship, though it is ascertained not to be the work of either Cyprian or Ambrose or Hilary or Augustine or Jerome,—but we point out where it is corrupt, giving warning in any case of flagrant error on the part of the translator, and explaining it, where the version is involved or obscure. If it is desirable that we should have the divine books as free from error in their text as possible, this labour of mine not only corrects the mistakes which are found in copies of the sacred volumes, but prevents their being corrupted in future; and if it is wished that they should be rightly understood, we have laid open more than six hundred passages, which up to this time have not been understood even by great theologians. This they admit themselves, as indeed they cannot deny it. If to that controversial theology, which is almost too prevalent in the schools, is to be added a knowledge of the original sources, it is to this result that our work especially leads. Therefore no kind of study is impeded by our labour, but all are aided.

Although we have translated throughout the reading of the Greek scribes, we still do not so approve it in every case, as not in some instances to prefer our own text, pointing out in every case, where the orthodox Latin writers agree or disagree with the Greek.[1]

'We do not tear up the Vulgate edition', Erasmus says in the letter quoted above. Is this really true? Does he not justify his new version (the translation from Greek) apart from a few exceptions?

[1] E.E. vol. III, no. 860 (32–61)=N. vol. III, pp. 430–1, of 26 August 1518.

THE PHILOLOGICAL VIEW: ERASMUS OF ROTTERDAM

Does he not intend his new version to be more easily understandable than the Vulgate? In his Preface to Valla's *Adnotationes* of 1505 he had pointed out that 'Valla collects his observations in a private commentary, and does not require you to change anything in your book'.[1] The same defence is taken up in his New Testament:

This is not written for the crowds, but for the erudite, and especially for the candidates of theology.[2] ... We have not called our work corrections but annotations. We are not irksome to anybody who quotes differently and who follows an old and corrupt edition.[3]

And before, in the first edition, he had written that the Vulgate and not his new version should 'be read in schools, sung in churches, quoted in meetings'. But, he continues, if his new translation is read at home, the official wording will be better understood.[4] Here we see Erasmus's aim: to further the understanding of the Bible. His work is not a substitute for the Vulgate but a help for the reader to understand the philosophy of Christ, for which no other philosophy is necessary. One has to approach it with 'a pious and open mind and particularly with a simple and pure faith'. For the Bible can be understood by everybody according to his ability.[5] Therefore Erasmus advocates renderings of the Bible into the vernacular languages. These passages of Erasmus's which were well known to Luther and Tyndale may be quoted in the translation by William Roy, an amanuensis of Tyndale.

And trulye I do greatly dissent from those men whiche wold not that the scripture of Christ shuld be translated in to all tonges that it might be reade diligently of the private and seculare men and women Other as though Christ had taught soch darke and insensible thinges that they

---

[1] *E.E.* vol. I, no. 182 (165–7) = *N.* vol. I, p. 383.
[2] Cf. *Apologia, N.T.1*, fol. bbb 6ʳ = *N.T.2*, p. 63: 'Et theologis potissimum hic meus desudavit labor....'
[3] *Contra Morosos*, no. 49, p. 74.
[4] *Apologia, N.T.1*, fol. bbb 7ʳ = *N.T.2*, p. 65.
[5] 'Parvis pusilla est, magnis plus quam maxima' (*Paraclesis, N.T.1*, fol. aaa 4ʳ⁻ᵛ = *N.T.2*, p. 8).

THE PHILOLOGICAL VIEW: ERASMUS OF ROTTERDAM

could scante be vnderstonde of a few divines. Or els as though the pithe and substance of the christen religion consisted chefly in this that it be not knowne...I wold desire that all women shuld reade the gospell and Paules epistles and I wold to god they were translated in to the tonges of all men So that they might not only be read and knowne of the scotes and yryshmen But also of the Turkes and sarracenes Truly it is one degre to good livinge yee the first (I had almoste sayde the cheffe) to have a litle sight in the scripture though it be but a grosse knowledge and not yet consummate) Be it in case that some wold laugh at it yee and that some shuld erre and be deceaved) I wold to god the plowman wold singe a texte of the scripture at his plowbeme And that the wever at his lowme with this wold drive away the tediousnes of tyme. I wold the wayfaringe man with this pastyme wold expelle the werynes of his iorney. And to be shorte I wold that all the communication of the christen shuld be of the scripture for in a maner soch are we oure selves as oure daylye tales are.[1]

The position of the Vulgate is thus left insecure: its authority only consists in its usage in church. This position had obviously to be reconsidered. At the fourth session of the Council of Trent on 8 April 1546, it was reaffirmed in a decree saying that 'the Vulgate, approved through long usage in the Church through so many centuries be held as authentic in public readings, disputations, preachings, and expositions, and that nobody dare or presume to reject it under any pretext'.[2] In the same decree the traditions,

---

[1] The translation is taken from: *An exhortation to the diligent studye of scripture made by Erasmus Roterodamus. And translated in to inglissh* (Hans Luft, Malborow, Hesse, i.e. J. Hochstraten, Antwerp, 1529, no pagination), leaf 5ᵛ–6ʳ. The text of Erasmus is found in *Paraclesis, N.T.1*, fol. aaa 4ᵛ=*N.T.2*, p. 8. Cf. Preface to *Adnotationes, N.T.1*, p. 229=*E.E.* vol. II, no. 373 (202 ff.). In 1535 Erasmus advocates that every theologian should be especially trained in the use of the vernacular. Though he generally only speaks of Italian (he refers to Dante and Petrarch), Spanish and French, he mentions the English and Saxon languages. Even the most barbarous language, he asserts, has its own elegance, emphasis, and grace (*Ecclesiastes sive de Ratione Concionandi*, II, L.B. vol. v, cols. 855–6). Cf. *E.E.* vol. IV, no. 1211 (277–9).

[2] 'insuper eadem sacrosancta synodus...statuit et declarat, ut haec ipsa vetus et vulgata editio, quae longo tot saeculorum usu in ipsa ecclesia probata est, in publicis lectionibus, disputationibus, praedicationibus et expositionibus pro authentica habeatur, et ut nemo illam reiicere quovis praetextu audeat vel praesumat' (*Canones et Decreta Concilii Tridentini*, ed. F. Schulte-A. L. Richter (Leipzig, 1853), p. 12). For origin and interpretation of this decree see P. Richard, 'Concile de Trente', in C. H. Hefele, *Histoire des Conciles d'après les documents originaux*, vol. IX, part I (Paris, 1930), and esp. H. Höpfl, 'Beitraege zur Geschichte der Sixto-Klementinischen Vulgata', *Biblische Studien*, vol. XVIII, parts 1–3 (Freiburg,

THE PHILOLOGICAL VIEW: ERASMUS OF ROTTERDAM

both the biblical and the oral, are reinstated. Their affirmation by the Council of Trent excludes the freedom of research as practised by the humanists. It is in keeping with these decrees that at the same Council parts of Erasmus's works were put on the *Index Librorum Prohibitorum*.[1]

It may be useful to recapitulate the different points which reveal Erasmus's attitude to the official version of the Bible. Though he does not wish to abolish the Vulgate for official usage, he believes that its meaning is not always clear. For the interpretation of the Bible it is necessary to go back to the language in the original version.[2] His own translation, based on the Greek original, makes the meaning clear and thus leads to the understanding not only of the Vulgate but also of the truth contained in the New Testament. Erasmus thus thinks that the official version cannot serve as a basis for interpretation. It has been shown in the first chapter that it is the essence of an authoritative version to replace the original in every respect. In Erasmus's view the Vulgate does not do so and cannot do so because of the deficiencies inherent in a translation. His edition thus directly challenges the medieval tradition of the Church, which was based on the Vulgate. But he does not wish to 'tear it up'. However, through his insistence that not the divine word but only the translation[3] is emended, the

---

Br., 1913), pp. 1 ff., esp. pp. 22 ff., where the above decree is interpreted as meaning that in questions of faith and doctrine the Vulgate is authentic. In all other questions the Vulgate is an ordinary translation only. 'Aus der Authentizitaet der Vulgata folgt ja nicht, dass jeder einzelne Satz in ihr ganz genau den Urtext wiedergibt...Buchstabenkritik ist nicht ihre Sache...' (pp. 25-6). But, he continues, the Vulgate has a special dignity which is found in no other translation, not even in the original language (p. 28). There was, however, a controversy about the meaning of this decree; especially the Spanish theologians thought that, according to it, the Vulgate is the only authentic version (Höpfl, *loc. cit.* pp. 30 ff.). Höpfl's interpretation is in agreement with the *Encyclica Providentissimus Deus* given by Pope Leo XIII of 18 November 1893 (*ibid.* p. 43).

[1] A clear and illuminating study on the trends of Catholic sixteenth-century writers is made by P. Polman, 'L'Élément historique dans la controverse religieuse du XVIe siècle', *Universitas Catholica Lovaniensis*, Series II, Tome 23 (Gembloux, 1932), see esp. pp. 284-309, where (p. 307 n. 1) some literature on the Council of Trent is referred to. The expurgated passages can be found in *L.B.* vol. x, cols. 1781-1844.

[2] For the views of Erasmus in his old age see, for example, *Ecclesiastes* (*L.B.* vol. v, col. 855).

[3] *Contra Morosos*, no. 11, p. 70.

THE PHILOLOGICAL VIEW: ERASMUS OF ROTTERDAM

meaning of the 'authoritative version' is considerably modified. An authoritative version no longer replaces the original but is used in church instead of the original text. For all questions of interpretation, however, the theologian has to refer to the original.

Erasmus turns away from the Latin Vulgate to the original language of the Scriptures and therefore necessitates reform within the Church or, as he would have preferred to say, a restoration of the Church to its 'pristine brightness'. It is, however, important to stress that he is sincere when he says that he does not wish to deviate from the judgement of the Church. He desires no break within the tradition, as all his writings testify. It is characteristic that in his *Ratio* he divides the Christian world into three circles: the clergy, free from worldly business, are nearest to Christ. They have to transmit the purity and the life of Christ to the second circle, namely to the princes, whose arms and laws serve Christ. The multitude is contained in the third circle, connected with God through the priests. Everybody has his fixed place within this order and cannot leave it but everybody must tend toward the first and highest circle; 'the boundaries between the circles are there for the purpose of a transformation, not towards the worse, but towards the better'.[1] For this transformation it is necessary to have a clear understanding of the Bible, especially the New Testament. The aim of the theologian is to narrate Scripture, to discourse about faith and piety, to move to tears, to inflame the mind to things divine (inflammare animos ad coelestia). Or, as he says at another place: One's aim should be 'to be changed, to be rapt away, to be inspired, to be transformed into what one learns'. One should pray to God for the help of the Holy Spirit.[2] This characterizes the religious feeling of Erasmus. The man who maintains that through learning and through understanding the literary meaning the comprehension of God's word is possible, who restates the old conception that

---

[1] *Ratio*, p. 27.
[2] *Ratio*, p. 15.

THE PHILOLOGICAL VIEW: ERASMUS OF ROTTERDAM

only knowledge of the letter can lead to a sound allegorical interpretation,[1] believes that man can be changed through learning. It is the rational approach to religious feeling that is taught by Erasmus. The Bible is the central point of a civilization which comprises the three circles of the Christian world. This catholicism of Erasmus's view leads him to oppose any attempt to reform Christianity outside the framework of the Catholic Church.

It was thus possible, and indeed natural, for Erasmus's New Testament to be praised by humanists who never thought of leaving the Catholic Church. There can be no doubt of the deep impression made by his edition on Thomas More, who wrote a long letter to a monk after the second edition had been published, using many of Erasmus's arguments in defence of his work. More does not believe that through changing the old translation Erasmus contradicts the Church. Like Erasmus he thinks that a new translation and interpretation will help to destroy the abuses current at his time and so help to restore Christianity.[2]

Colet after receiving the New Testament writes to Erasmus: 'the name of Erasmus will never perish.' He admits that with the help of Erasmus's work he understands passages which he had not been able to understand before. He therefore decides to learn Greek 'without which we are nothing'.[3]

Both Colet and Thomas More mention that there is opposition to Erasmus's New Testament from people 'whose censure is praise and whose praise is censure', as Colet writes.[4] But More points out that he is perturbed, since these opponents are very powerful.[5] The writings of these opponents of Erasmus constitute the second

[1] Preface to the *Adnotationes*, N.T.1, p. 227.
[2] *Mo.E.* no. 83 *passim*, esp. ll. 680 ff., 692 ff., 1005, 1061 ff.; ll. 983–1009 are on Englishmen praising Erasmus's New Testament. Some references to this praise are found in J. A. Faulkner, *Erasmus: The Scholar* (New York, 1923), pp. 124–6.
[3] *E.E.* vol. II, no. 423 (5 ff., 47–8); *E.E.* vol. II, no. 471 and Thomas More's letter to Erasmus of 22 September ⟨1516⟩ with an interesting characteristic of Colet (*E.E.* vol. II, no. 468 (11–17)).
[4] *E.E.* vol. II, no. 423 (7–11).
[5] *E.E.* vol. II, no. 481 (31 ff.), of 31 October ⟨1516⟩. Cf. P. S. Allen, *Erasmus, Lectures and Wayfaring Sketches*, p. 85.

THE PHILOLOGICAL VIEW: ERASMUS OF ROTTERDAM

great battle against humanism in the sixteenth century. Erasmus answered all the pamphlets directed against him in *Apologiae*.[1] It is not the purpose of this study to enter into the details of this fight, which is often poisoned by personal remarks that reveal the weakness and meanness of an adversary. All these attacks are directed against Erasmus's Latin version, not against the publication of the Greek text. Erasmus is reproached for having mistranslated certain words or phrases. Important though these censures are, they may be passed over in this study.[2] Only the general tendencies that guide these attacks need be described.

It was in 1514 that Martin Dorp,[3] on the instigation of the theologians at the university of Louvain, wrote to Erasmus asking him to desist from his work on the New Testament. Thus almost two years before the publication of his Bible he received the first warning. Dorp wrote a second letter in 1515 restating his arguments, although in 1516 he became convinced of the correctness of Erasmus's views.[4] Dorp makes the following points. If the Vulgate contained falsifications of the original Bible and errors, the Church would have been wrong for many centuries.[5] This is

---

[1] These are published in *L.B.* vols. IX and X.
[2] Erasmus's accuracy or want of it in his edition which, through Stephanus's edition, Paris, 1550, influenced the Textus Receptus (F. G. Kenyon, *Handbook to the Textual Criticism of the New Testament*, 2nd ed. London, 1926, pp. 268–71) will not be discussed here. A detailed criticism of his text of the Revelation is given by Franz Delitzsch, *Handschriftliche Funde*, 2 vols. (Leipzig, 1861, 1862). For further textual criticism of his text see the prefaces to Greek editions of the New Testament; cf. E. Nestle, *Introduction to the Textual Criticism of the Greek New Testament*, translated by W. Edie (London, 1901), pp. 3–5, and the literature mentioned by P. S. Allen, *E.E.* vol. II, no. 373 and the introduction to this letter (p. 166). The story of the controversy about the New Testament is described in most biographies of Erasmus. A special study on all the different phases is A. Bludau, 'Die beiden ersten Erasmus-Ausgaben des Neuen Testaments und ihre Gegner', *Biblische Studien*, vol. VII (Freiburg im Br., 1902). Cf. R. Simon, *Histoire critique des principaux commentateurs du Nouveau Testament* (Rotterdam, 1693), pp. 484–536. The controversy between Erasmus and Jacques Lefèvre is dealt with by M. Mann, *Érasme et les débuts de la Réforme Française (1517–1536)*, Bibliothèque Littéraire de la Renaissance, N.S. 22 (Paris, 1934). Erasmus and Zúñiga (Stunica): M. Bataillon, *Érasme et l'Espagne, recherches sur l'histoire spirituelle du XVIe siècle* (Paris, 1937), pp. 98–105. P. S. Allen, in *E.E.* vol. IV, Appendix XV, pp. 621–2.
[3] For the relationship between Erasmus and Dorp see O. Hendriks, *Erasmus en Leuven* (Bussum, 1946).
[4] *E.E.* vol. II, nos. 304 of ⟨c. September 1514⟩ and 347 of 27 August 1515.
[5] *E.E.* vol. II, no. 304 (92–110).

THE PHILOLOGICAL VIEW: ERASMUS OF ROTTERDAM

impossible; the references of most Councils of the Church to the Vulgate leave no doubt that the Church considers this Latin version, and not the Greek original, as the official Bible, and that the Vulgate is authorized,[1] a view opposed by Erasmus.[2] The logical conclusions from this argument are that the Vulgate must be without mistakes. Discrepancies between the Greek and Latin texts are due to changes of the wording by the schismatic Greeks.[3] If Erasmus believes that he can, without changing the meaning through emendation, arrive at a more significant reading, he may do so. If, however, he has to change the meaning, he should give the reasons thereof in a special appendix.[4]

Much more outspoken and more venomous is Diego Lopez Zúñiga (Stunica), a scholar who, knowing the three holy languages, was one of the editors of the famous Complutensian Bible. Zúñiga sums up Erasmus's New Testament thus: It is an open condemnation of the version of the Church. If Erasmus, relying on his knowledge of secular literature, had not been vainglorious, he would not have made a new translation but would have written notes only, where, during the century-long tradition, the text had become corrupt.[5] Dorp and Zúñiga, who insist on the preservation of the authoritative version, have clearly recognized that it cannot be admitted that grave mistakes are contained in it, for otherwise no authority is left to it, an argument taken from St Augustine.[6] Besides attacking Erasmus's view that the language of the Bible should be more elegant than that of the Vulgate[7] these two scholars, believing as they did that the Vulgate was the authoritative version, drew the logical conclusion that Greek is not necessary for its understanding.[8]

---

[1] *E.E.* vol. II, no. 347 (209–37).
[2] *E.E.* vol. II, no. 337 (771 ff.).
[3] *E.E.* vol. II, no. 304 (110–18); no. 347 (170–87).
[4] *E.E.* vol. II, no. 304 (141–6); no. 347 (190–200).
[5] *Annotationes Iacobi Lopidis Stunicae contra Erasmum Roterodamum in defensionem tralationis Novi Testamenti* (Alcalá, 1519), Prologus, fol. A iᵛ.
[6] *P.L.* vol. XXXIII, cols. 112–13. *E.E.* vol. II, no. 304 (138–40).
[7] Stunica: *loc. cit.*; Dorp: *E.E.* vol. II, no. 347 (213–20).
[8] *E.E.* vol. II, no. 347 (337–49).

THE PHILOLOGICAL VIEW: ERASMUS OF ROTTERDAM

Jacob Latomus[1] also advocated the view that Greek is unnecessary for the understanding of the New Testament. 'God', he maintained, 'has not bound together His wisdom and His law with certain letters or apexes of any individual language.'[2] From this passage Latomus developed the argument against Erasmus's method. The Arians could not understand the Bible in spite of their knowledge of Greek. Nobody is able to read the Gospel better than the Church. 'The evangelical law is not led by the will of pope, council or Church, it leads them, it is not ruled but it rules, it does not follow, it procedes.'[3] For the pious the knowledge of languages is not necessary, for the impious it is of no value. As the Gospel leads to piety and as the Gospel can be understood by pious people only, Latomus arrives at a conclusion which he himself calls absurd, namely that those who are pious are in no need of Holy Scripture, not to speak of languages, but for those who are not pious Holy Scripture is useless.[4] Only one more argument of Latomus need be added: God, who at the beginning instructed the Church, does not desert it and thus the Church, God's mystical body, preserves the tradition which is found in the writings of the Schoolmen.[5]

Latomus criticizes Erasmus's view at every point. It is an attempt to re-establish the weight of authoritative thought against the rise of individualism. It is the upholding of the mystical tradition of the Church as God's body against rational disintegration. It is the wish to preserve the trends of thought prevalent at the beginning of the sixteenth century which were attacked by the new doctrine and method promulgated by Erasmus.

---

[1] *De Trium Linguarum et Studii Theologici Ratione Dialogus* (sine loco et anno) (Froben, Basel, 1518). A new edition has been published by F. Pijper, in *Bibliotheca Reformatoria Neerlandica*, vol. III ('s-Gravenhage, 1905), pp. 43-84. Unfortunately the numbers of the paragraphs quoted above (which are taken from the first edition) are not printed in the new edition (to which the page numbers in brackets refer). Erasmus calls Latomus's book 'very erudite and elegant' (*E.E.* vol. III, no. 934 (3), of 30 March 1519, where also see for biographical details on Latomus).

[2] Latomus, *op. cit.* no. 40 (p. 56).
[3] Latomus, *op. cit.* no. 41 (p. 57).
[4] Latomus, *op. cit.* no. 41 (p. 58).
[5] Latomus, *op. cit.* nos. 63 ff. (pp. 65 ff.).

THE PHILOLOGICAL VIEW: ERASMUS OF ROTTERDAM

All the criticisms discussed above have as a common denominator the preservation of the authority of the Church against innovation. Erasmus and his opponents have one common basis, namely, the belief in the Church, even though abuses had temporarily taken hold within it. There was, however, slowly developing another criticism which did not share this common platform and was ready to leave the Church. Men who followed this view could, in their censure, be less careful than was Erasmus. It was inevitable that they should turn against Erasmus's method of Bible interpretation and against his humanistic views. These were the men of the Reformation.

CHAPTER VI

THE INSPIRATIONAL VIEW: LUTHER

For the appreciation of Luther's translation of the Bible it is of the utmost importance to see how his religious feeling and experience caused him to react to the humanistic tendencies of his time. Though it is difficult to assess the influence of humanism on the young Luther, one fact stands out clearly.[1] Without the foundation laid by Reuchlin, Erasmus and other humanists, Luther would not have been able to pursue Greek and Hebrew studies in the way he did. It is, moreover, well to remember that he was always in contact with Johannes Lang, who knew Greek[2] and Hebrew, and that the humanist Melanchthon was in Wittenberg from 1518 onwards. Though it is well known how great Luther's hold was on this scholar,[3] it is easy to forget that Luther in turn was influenced by Melanchthon.

The early experiences of Luther's religious life are comparatively well known. One event only, which has a direct bearing on the subject of this study, need be discussed here. Unfortunately, its exact date is unknown and the longest report on it by Luther was not written until 1545, more than thirty years after it had taken place. Luther's memory of this event may have been unreliable, and it is therefore necessary to exercise great caution before considering this later report as a primary source. But it may legitimately be used for finding out Luther's judgement on this important event of his life. Fortunately there are earlier references to it by Luther which seem to suggest that his original evaluation of it in no way contradicts the later version but actually confirms

---

[1] *L.E.* vol. I, no. 60 (1–16), of 19 February 1518. There is a cautious and well-balanced account of the humanistic influence on Luther at Erfurt by H. von Schubert, 'Luthers Frühentwicklung bis 1517/9', *Schriften des Vereins für Reformationsgeschichte*, vol. XXXIV (1916), pp. 11–12.

[2] *L.E.* vol. I, no. 14 (21–2), of 29 May 1516.

[3] K. Hartfelder, 'Philipp Melanchthon als Praeceptor Germaniae', *Monumenta Germaniae Paedagogica*, vol. VII (Berlin, 1889), pp. 204–7.

THE INSPIRATIONAL VIEW: LUTHER

it in every point. Some of these earlier statements were written before the translation of the Bible and are thus of great significance for this study. They can be more easily understood if they are set against Luther's biographical remarks of 1545.

In 1545 Luther relates how in his youth he hated the word 'righteousness' since he understood it philosophically in the meaning that God is righteous and punishes the sinners and unrighteous. He found it very difficult to understand the connexion of the first and second parts of Romans i. 17 which reads: 'For therein is the righteousness of God revealed from faith to faith: as it is written, The just shall live by faith.' For days and nights he meditated over the meaning of this verse. Then he began to perceive that the word 'righteousness' means that God in His mercy justifies man by faith.

At this point [Luther continues] I felt completely reborn and as if I had entered paradise with its open gate. Forthwith the aspect of the whole of Scripture seems to have changed. Thereafter I ran through the Scriptures as I had them in memory and collected analogical meanings in other words, such as the work of God, that means the work that God works in us, the virtue (virtus) of God, that means the virtue through which He makes us powerful, the wisdom of God, that means, the wisdom through which He makes us wise, the courage of God, the salvation of God, the glory of God. My love for that sweetest word 'righteousness of God' was henceforth as great as my hatred for it had been hitherto. In this way this passage of Paul was truly the gate of Paradise.

Later, Luther concludes, he found a confirmation of this view in the works of Augustine.[1]

---

[1] This passage is found in the Preface to an edition of his collected works, *W.A.* vol. LIV, 185 (12)–186 (20), esp. 186 (8–16)=Scheel, pp. 191 (28)–192 (27). H. Denifle, *Luther und Luthertum in der ersten Entwicklung planmässig dargestellt*, vols. I, II (Mainz, 1904, 1909), has tried to prove that Luther's interpretation of Rom. i. 17 had been known throughout the Middle Ages. He has collected all the sources in *Ergänzungen zu Denifles Luther und Luthertum, 1, Quellenbelege. Die abendländischen Schriftausleger bis Luther über Justitia Dei (Rom. 1, 17) und Justificatio* (Mainz, 1905). Luther himself says that after his discovery of the meaning of Rom. i. 17 he found his exegesis in Augustine (Preface, *ibid.*). There is a special paper on this subject from the protestant point of view by K. Holl, 'Die iustitia dei in der vorlutherischen Bibelauslegung des Abendlandes' (1921), reprinted in *Gesammelte Aufsätze zur Kirchengeschichte*, vol. III (1928), pp. 171–88. Discussions on this subject are found in almost every book on Luther.

THE INSPIRATIONAL VIEW: LUTHER

This religious experience was the birth of Luther's doctrine of 'justification by faith'.[1] Luther himself, in 1542 and in his *Table Talk* of 1532,[2] ascribed it to 'illuminatio'. This discovery of the meaning of Romans i. 17 will be called Luther's 'inspiration' in the following pages in which it will be attempted to prove that his interpretation of his 'inspiration' did not materially change during his life.

In 1545 Luther said that this recognition changed the whole Bible for him and that he re-interpreted other words in the light of his experience. The process described is clear. The meaning of one passage had been revealed to him. He therefore had received the true understanding of this one verse. Holy Writ, being God's revelation, must of necessity be a unity and its contents be in agreement. It is therefore permissible, or even necessary, to interpret the Bible in accordance with Romans i. 17, if the true meaning of this verse has been revealed.[3] This is the more possible as Paul himself interprets Habakkuk's phrase (ii. 4) 'The just shall live by faith' in Romans i. 17. It is important to note that St Paul is here expounding a sentence of the Old Testament. When Luther came to understand the connexion between the two parts of Romans i. 17, he believed that he had not only perceived the true meaning of St Paul but also his method of interpreting the Old Testament. It is a direct consequence of his inspiration that he interpreted the whole Bible in accordance with St Paul.

If Luther indeed believed that his inspiration enabled him to understand the Bible, he could justifiably say that God had given him the true comprehension of Holy Writ: 'I have not dared nor am I able to boast of anything but of the word of truth which the

---

[1] For details see, for example, A. Harnack, *Lehrbuch der Dogmengeschichte* (4th ed. Tübingen, 1910), vol. III, pp. 820ff., 843ff.; R. Seeberg, *Lehrbuch der Dogmengeschichte* (4th ed. Leipzig, 1933), vol. IV, 1, pp. 129ff.

[2] *Ennarratio in Genesim* of 1542, to Gen. xxvii. 38, *W.A.* vol. XLIII, 537 (21–6) = Scheel, p. 167 (9–13); *T.R.* vol. II, no. 1681; *T.R.* vol. III, no. 3232a of 1532 = Scheel, pp. 91 (7–34), and 94 (1–10).

[3] Cf., for example, Luther's *Disputatio Iohannis Eccii et Martini Lutheri Lipsiae habita* of 1519, *W.A.* vol. II, p. 302 (1): '...verbi intelligentia ex tota scriptura...petenda est...'; p. 361 (19ff.): 'Oportet ergo theologum si nolit errare, universam scripturam ob oculos ponere...'. The unity of the Bible is assumed throughout the Middle Ages, see F. Kropatscheck, *Das Schriftprinzip der lutherischen Kirche*, vol. I (Leipzig, 1904), p. 426.

THE INSPIRATIONAL VIEW: LUTHER

Lord has given me.'[1] Whilst these words of 1521 may leave some doubt as to their real meaning, Luther's view cannot be misunderstood in the following passage found in a letter of 1522:

> Your Excellency knows, or if not may be informed herewith that I have received the Gospels not from man but solely from heaven through our Lord Jesus Christ (Gal. i. 10ff.) and that I could have prided myself in words written and spoken a servant of God and his Evangelist, and this I will do henceforth.[2]

It cannot be expected that Luther should make public statements of this kind which could easily expose him to bitter attacks.

'I have not received the Gospel from man but from heaven.' When he used this phrase of St Paul, did Luther mean that he had a perfect understanding of every passage of the Bible? He never claimed this. He meant, I assume, that he had found the key which opened up the meaning of the Bible as a whole but not of every sentence in it. Luther wrote in 1545 that he had reinterpreted single words by analogy with the term 'righteousness', but this does not mean that he was able to expound the full significance of every passage. Even his discovery that every sentence of Holy Writ speaks either of grace or law did not help him to understand it at once. It is only through God's grace that the full meaning of any passage can be found. This seems to be the meaning of a sentence written in 1519 which, I venture to suggest, expresses Luther's own view on his inspiration. Speaking of the Psalms he says:

> It is sufficient to have understood some and these partly, the spirit reserves much to himself that he may thus hold us as his pupils for ever, he shows us much for the sole purpose of attraction, he makes known much that he may excite our emotion.[3]

It is, in Luther's view, never possible to be satisfied with any interpretation. It is necessary to receive God's grace anew for the

---

[1] L.E. vol. II, no. 429 (14–16), to Spalatin, of September 1521.
[2] L.E. vol. II, no. 455 (39–43), to Kurfürst Friedrich, of 5 March 1522. 'E.K.F.G. weisz, oder weisz sie es nicht, so lasz sie es ihr hiemit kund sein, dasz ich das Euangelium nicht von Menschen, sondern allein vom Himmel durch unsern Herrn Jesum Christum habe (Gal. i. 10ff.), dasz ich mich wohl hätte mügen (wie ich denn hinfort tun will), einen Knecht und Euangelisten rühmen und schreiben.' Cf. L.E. vol. II, nos. 456 (50–4), 464.
[3] *Operationes in Psalmos*, Dedication, dated 27 March 1519 (W.A. vol. V, p. 22 (28–30)).

THE INSPIRATIONAL VIEW: LUTHER

interpretation of every passage. Therefore Luther advised his friends to pray to God to grant them the full recognition of His word. Testimonies of this kind abound in Luther's works. Here is one:

> Therefore the first duty is to begin with a prayer of such a nature that God in His great mercy may grant you the true understanding of His words....[1]

And another of 1521:

> Nobody can understand God or God's word unless he receive it directly from the Holy Spirit.[2]

These quotations, dating from 1518 onwards, seem to prove that Luther's inspiration was the central point from which his interpretation of the Bible was derived. He always believed that every verse of Scripture was bound to agree with his own illumination. Therefore all his interpretations of theologically important words are not only in harmony with one another, but they also reflect the impact of his religious experience. The student observes that Luther's exegesis never contradicts his inspiration. This remarkable fact will be illustrated in the following pages. It should, however, not be imagined that Luther was able to perceive the full results of his inspiration at once. This experience had, as he himself said, to be applied first of all to the exegesis of analogous words. Then he could proceed to a fuller interpretation of the Bible and to a criticism of the existing schools of thought. All this took years. He could not, at the time of his inspiration, perceive that his religious experience would lead him to a break with the Roman Catholic Church, nor could he imagine that he would have to abandon the fourfold interpretation of Scripture as practised by the Schoolmen. There must, of necessity, have been a period in Luther's life when scholastic views and the ideas derived from his inspiration are found side by side. Luther naturally endeavoured to use the traditional method which he had

---

[1] L.E. vol. I, no. 57 (32–5), to Spalatin, of 18 January 1518.
[2] W.A. vol. VII, p. 546 (21–9): 'Denn es mag niemant got noch gottes wort recht vorstehen, er habs denn on mittel von dem heyligen geyst.'

learned at the university, and to express his own experience in this way. This can be seen from his *Lecture on the Psalms* of 1513–15.[1]

When in 1516 Erasmus's New Testament was published Luther's inspiration furnished him with a firm basis from which to view this work of humanistic scholarship. Erasmus's publication provides the student with a convenient landmark for dealing with Luther's own Bible interpretation. There is some difference in his technique before 1516 and after. Before, he was feeling his way without, it seems, being conscious of his final aim. Afterwards, he is surer of the method to be applied and this helps him to abandon the traditional medieval way of exegesis. But his earlier work must be discussed before the significance of the humanistic commentaries for his method can be fully recognized.

In the *Lecture on the Psalms* of 1513–15 Luther postulates that all Psalms refer to Christ and must be interpreted in the light of this view, which, of course, is traditional. It is fortunately possible to state that Luther's immediate source for this proposition was Jacques Lefèvre's *Quincuplex Psalterium*,[2] a book first published in 1509. Luther's copy of Lefèvre's work has been preserved, and his notes in it bear witness to the fact that he used it for the preparation of his *Lecture on the Psalms*.[3] It is important not to forget

[1] It is not clear when the different parts of this lecture were written; cf. E. Hirsch, 'Initium theologiae Lutheri', *Festgabe für D. Dr. Julius Kaftan...zu seinem 70. Geburtstage* (Tübingen, 1920), pp. 161 ff., and H. Böhmer, 'Luthers erste Vorlesung', *Berichte der Sächsischen Akademie der Wissenschaften* (Leipzig, Philol.-hist. Kl. 75, 1923), pp. 19 ff., esp. p. 29. H. Böhmer assumes that some parts were written in 1516 only. For a discussion on this issue see J. Mackinnon, *Luther and the Reformation*, vol. I (London, 1925), pp. 147 ff.; O. Scheel, 'Luthers Rückblick auf seine Bekehrung in der Praefatio zu seinen gesammelten Schriften', *Zeitschrift für Theologie und Kirche*, vol. XXI (1911), pp. 89–122.

[2] The importance of Lefèvre for Luther has been mentioned by N. Weiss, *Société du Protestantisme Français, Bulletin Historique et Littéraire*, vol. XXXVII, no. 188, p. 163 n. 3, repeated by M. Mann, 'Érasme et les débuts de la Réforme Française (1517–1536)', *Bibliothèque Littéraire de la Renaissance*, n.s., vol. XXII (Paris, 1934), p. 14 n. 1. Another short note to this effect is found in Böhmer, *op. cit.* p. 20, and in the preface to Kawerau's edition of Luther's notes (*W.A.* vol. IV, pp. 463–4). E. Vogelsang, 'Die Anfänge von Luthers Christologie nach der ersten Psalmenvorlesung', *Arbeiten zur Kirchengeschichte*, vol. XV (1929), pp. 21–2. A. Hamel, *Der junge Luther und Augustin* (Gütersloh, 1934), traces Augustine's influence on Luther's *Lecture*.

[3] These notes are printed in *W.A.* vol. IV, pp. 466–526. The *Lecture on the Psalms* is printed *W.A.* vols. III–IV. A short analysis of Luther's view is found, for example, in

THE INSPIRATIONAL VIEW: LUTHER

Luther's dependence on earlier writers, even where his ideas may seem to be in close harmony with the outcome of his inspiration. Indeed, it has been observed that Luther refers to 'justification by faith' in his *Lecture* and that therefore the inspiration took place either before 1513 or between 1513 and 1515 when he actually wrote it. Yet it is necessary to compare Lefèvre's method and his fundamental although traditional idea that the Psalms refer to Christ with Luther's thoughts as expressed in his commentaries.

In his Preface to the *Quincuplex Psalterium* Lefèvre analyses[1] especially the meaning of the term 'literal sense'. In this term he discovers two different contents. The one that is commonly used interprets the Psalms as if these divine hymns of David refer to his persecution by Saul and to his other wars. This, Lefèvre maintains, is the exegesis of the Jews and the Rabbis whom 'God has smitten with blindness' and who therefore are unable to understand the divine. This interpretation does not lead to real understanding; those who follow its method turn away from the Psalms, sad and dejected. 'Far be it from us to believe that this is the real content of the term "literal sense"—(the so-called sense of the letter) and to make David a historian rather than a prophet.'[2] He then tries to lay the foundation of a different explanation of the term 'literal sense'. The apostles and prophets, he says, have shown the way which leads to the understanding of this term. They have shown that for the interpretation of the Psalms it is necessary to find out 'the intention of the prophet [David] and of the Holy Spirit speaking in him'.[3] Quoting some examples where

R. Seeberg's *Lehrbuch der Dogmengeschichte*, vol. IV, no. 1, pp. 83 ff. The following books were of value to me: H. Preuss, *Die Entwicklung des Schriftprinzips bei Luther bis zur Leipziger Disputation* (Leipzig, 1901); K. Holl, 'Luthers Bedeutung für den Fortschritt der Auslegungskunst' (1920), reprinted in *Gesammelte Aufsätze zur Kirchengeschichte*, vol. I (1927), pp. 544–82; H. Böhmer, *op. cit.* pp. 1–28. Böhmer justly criticizes Kawerau's edition in *W.A.* vols. III–IV indicating the difficulties of using it. The omission of quotations of the Bible in *W.A.* vols. III–IV is criticized by K. A. Meissinger, *Luthers Exegese in der Frühzeit*, Dissertation (Giessen, 1910), pp. 22–6, 29–36; see *ibid.* pp. 7–14 for corrections of Kawerau's readings and emendations of Luther's text.

[1] For Lefèvre's method see K. H. Graf, 'Jacobus Faber Stapulensis', *Zeitschrift für die historische Theologie*, vol. XXII (1852), pp. 25 ff., and M. Mann, *op. cit.* pp. 13 f.
[2] Lefèvre, *Quincuplex Psalterium* (H. Stephanus, Paris, 31 July 1509), fol. aᵛ.
[3] *Ibid.* fol. aʳ.

the apostles, and especially Paul, have referred to the Psalms, Lefèvre comes to the conclusion that the Psalms speak of Christ and not of David. 'We call the literal sense', he writes, 'that which tallies with the Spirit and which is revealed by the Holy Spirit' or, as he also says, which 'is infused by the divine Spirit'. Lefèvre argues that according to St Paul's phrase: 'we know that the law is spiritual' (Romans vii. 14), the literal sense of the law is spiritual and therefore the literal and spiritual senses coincide. This single, and new, meaning is that 'which the Holy Ghost intends when speaking in the prophet'. It is this sense which Lefèvre attempts to elicit in his *Quincuplex Psalterium*, 'as much as the Spirit of God gave to us'.[1] From the belief that the Psalms are prophetic, Lefèvre develops a second fundamental assumption, namely that the true contents of the prophetic meaning can be fully understood only by an expositor who is himself a prophet. But Lefèvre points out that he is no prophet, and that therefore his interpretation is open to criticism.

This preface is an attempt to introduce a new method of exegesis which enabled Lefèvre to break away from the traditional method of the fourfold interpretation of the Bible. The literal sense which used to be the basis of the other interpretations plays no important part in his plan, since it is replaced by the prophetic sense. This prophetic sense can be understood only by a prophet who is himself inspired by the Holy Ghost as were the prophets of old. Therefore, Lefèvre, being no prophet, intends to follow those who have interpreted the Psalms by inspiration and who have believed that they refer to Christ.

Luther, as mentioned above, also believes that all the Psalms speak of Christ, but at this stage of his development he was not yet prepared to break with tradition. He used interlinear and marginal glosses for the explanations of single words, for references and short notes, and *scholia* for discussions of the contents and for digressions. These digressions may be of considerable length and may deal with subjects that have no conceivable con-

[1] Lefèvre, *Quincuplex Psalterium*, fol. a^v.

THE INSPIRATIONAL VIEW: LUTHER

nexion with the text. For example, Luther includes a complete sermon on the Martyrs in his discussion on Psalm lix (lx).[1] The threefold notes gave Luther an opportunity to speak about the prophetic meaning of the Psalms. This thought is, of course, identical with Lefèvre's. In the prefaces he, like his predecessor, refers to the Jews who interpreted the Psalms as if they bore upon history and not upon Christ. But Christ 'opened the understanding' of those who believed in Him, 'that they might understand the Scriptures'.[2] This quotation of Luke xxiv. 45 embodies an important principle of Luther's which may be said to have two aspects: first, the understanding of Scripture is possible only to Christians; and secondly (if the word 'He *opened* the understanding' is stressed), a revelation is necessary for him who wishes to understand the words of the prophet David. Like Lefèvre, Luther has two meanings in mind. The Jews cannot understand that 'every prophecy and every prophet must be understood to be concerned with Christ unless it is made perfectly clear that the words are spoken about somebody else...'. From this he draws the same conclusion as Lefèvre, that all the Psalms must be understood in a prophetic sense, and this at once leads him to a new attack on the Jews and on all those who in the past have not interpreted the Psalms after his own fashion. These commentators obviously followed the Jewish exegesis. Here are his words:

Those who do not expound a great number of Psalms prophetically but historically follow some Hebrew rabbis who are writers of deceptions and fabricators of Jewish falsehoods.[3]

---

[1] This sermon is *W.A.* vol. III, pp. 342 (38)–346 (53). Luther had special books printed with ample space for both interlinear and marginal glosses. Facsimiles of his lectures have been published: *Auslegung des Römerbriefes 1515–1516. Autograph der königlichen Bibliothek zu Berlin*, MS. theol. lat. qu. 21 (1909). *Luthers Vorlesungen über den Galaterbrief 1516/17* ed. H. von Schubert, *A.H.A.* 5. Abh. (Heidelberg, 1918). The best description of Luther's way of making notes is in J. Ficker's edition, *Luthers Vorlesung über den Römerbrief 1515/1516* (4th ed., Leipzig, 1930) (*Anfänge reformatorischer Bibelauslegung*, vol. I), pp. xlix–li. Cf. H. von Schubert-K. Meissinger, 'Zu Luthers Vorlesungstätigkeit', *S.H.A.* (1920), 9. Abh., pp. 22ff.   [2] *W.A.* vol. III, p. 11 (15–16).

[3] *W.A.* vol. III, p. 13 (6–7, 9–11). The word 'fabricator' in the last sentence is a free rendering of 'figuli' ('potters').

THE INSPIRATIONAL VIEW: LUTHER

Luther goes one step further. He is in no way aware of the importance of Hebrew for the interpretation. He even rejects those who used the Hebrew text in support of their commentaries. It leads to suspicion, Luther asserts, and it is not safe to listen to those who mention the Hebrew literal 'truth' if they use it not for the illumination of Christian 'truth', but only for its condemnation. A few sentences later he reproaches Nicolaus of Lyra and even more Paul of Burgos for preferring the reading of the Hebrew text to that of the Vulgate.

But all this seems to be said rather according to the Jewish conjecture, for in their carnal understanding they adjust the text of Scripture as it seems good to their minds [sensus].[1]

Luther was thus not prepared to interpret the biblical text according to the wording of the original languages.

The rejection of the 'carnal understanding' in favour of the spiritual meaning causes Luther to speak about the inspiration which the commentator of a prophet should possess. After having said a few words about the 'most illustrious prophet David' he admits his inability to interpret the Psalms:

I candidly admit that I have not yet understood a very great number of Psalms and will be unable to interpret them unless the Lord illuminate me for the sake of your merits as I trust.[2]

These words are not merely a declaration of modesty, they must be understood in their context. A few sentences later quoting among others Psalm lxxii. 16–17 (lxxiii. 16–17), Mark ix. 34 and Amos vii. 14, he maintains that only he who 'goes into the sanctuary of God' and understands 'concerning their last ends'[3] is fully able to perceive their real meaning.

I undertake [he writes] to interpret a prophet though certainly 'I am not a prophet nor a prophet's son' [Amos vii. 14].[4]

---

[1] *W.A.* vol. III, p. 518 (16–30) on Psalm lxxiv. 9 (lxxv. 8): '...carnaliter...intelligentes...'. Cf. Luther's Sermon of 26 December 1514 where he explains: 'Est enim sapientia carnis, quae vulgo significatur Sinnligkeit, Sensualitas...' (*W.A.* vol. I, p. 34 (1–2); vol. III, p. 11 (14f.)).   [2] *W.A.* vol. III, p. 14 (4–6).

[3] This translation is taken from the Douay-Rheims version.

[4] *W.A.* vol. III, p. 14 (4–6, 16–17); cf. Böhmer, *op. cit.* p. 48.

THE INSPIRATIONAL VIEW: LUTHER

The passages quoted are without doubt reflections of Lefèvre's views—all the elements of which reappear in Luther's work: the rejection of what is generally called 'literal sense' in favour of the prophetic contents which refer to Christ, and the idea that only a prophet is able to interpret prophecies. Luther, however, went further by using Paul's terminology for the exegesis of the Psalms. As shown above (p. 169) this tendency is the direct outcome of his inspiration. For this reason alone the atmosphere of the two prefaces differs. Although Lefèvre quotes the Bible fairly frequently, Luther abounds in quotations and allusions to passages of Holy Writ. For example, Lefèvre mentions two lines of Homer when saying that he is no prophet, whilst Luther quotes Amos.

Apart from Lefèvre, Luther used the usual commentaries such as Augustine, the ordinary gloss, Peter Lombard, but also Reuchlin,[1] that is to say, he tried to profit from tradition and even from the latest research. But the advantages of the fourfold method of interpretation were greatly reduced by his substituting the prophetic sense for the strictly literal meaning. Luther relied on the text of the Vulgate although in the later parts he became doubtful of its accuracy. Lefèvre had printed Jerome's version from Hebrew in the third column of his *Quincuplex Psalterium*, calling it 'Hebraeus', a name taken over by Luther.

If Luther's negative and hostile attitude towards the Hebrew commentators is remembered, the reader of his *Lecture* may well find it strange that he called in Hebrew grammar to help him in his exegesis. There is, it would seem, another reason why Luther should not make use of Hebrew. His principle of treating the literal meaning as spiritual may sometimes be incompatible with the method of grammatical analysis of the Hebrew text as practised by Reuchlin. Was Luther aware of these contradictions? Did he perhaps believe that grammatical and spiritual interpretations could always work peacefully together? It seems to me that he did, for he could not foresee that the spiritual exegesis would or could conflict with the grammar. Between 1513 and 1515 Luther

---

[1] Böhmer, *op. cit.* p. 23 n. 1.

did not know Hebrew well enough; even in 1518 he could not yet understand the original text of the Psalms.¹

It is interesting to observe how Luther made use of Hebrew in his *Lecture on the Psalms*. He learned that in Psalm ii. 10 'intelligite' (be wise) had been used to translate the Hebrew causative form 'Hiph^eil' ('Be made wise').² By analogy he interpreted 'scitote' in Psalm iv. 4 as a causative form, although no causative verb form is found in the Hebrew text. From Reuchlin's grammatical explanation he drew the conclusion that the meaning of Psalm ii. 10 is that Christ makes man wise. This is in contrast to Reuchlin who maintained that erudition makes man wise. Luther's belief that man is made wise by God is in conformity with the report he made in 1545 on his inspiration. There Luther writes: the word 'righteousness' means to be made righteous by God; by analogy the words of the Psalm 'be wise' mean 'to be made wise by God'. Whatever the exact date of Luther's interpretation of Psalm ii. 10, its meaning fully agrees with his inspiration. But such is the complexity of human thought that it is impossible to conclude from this alone that Luther's inspiration must of necessity have taken place before he wrote these words. Luther knew, of course, the ordinary gloss to this Psalm, and there he found a passage bearing Augustine's name to the effect that kings are 'under Him who is the giver of intellect and erudition'.³

This example gives an insight into Luther's work on, and method of, the exegesis in his *Lecture on the Psalms*. Clearly, he did not arrive at his result from the study of Hebrew, for there is no causative form in the Hebrew text of Psalm iv. 4. Luther saw

---

¹ Lefèvre did not know Hebrew, as Mutianus wrote in 1514 (*Mu.E.* no. 387, p. 47). Luther quotes Reuchlin, for example, in his commentary to Psalm cxxxviii. 14 (cxxxix. 15); cf. C. H. Graf, 'Jacobus Faber Stapulensis', *loc. cit.* p. 22. For Luther's slow progress in learning Hebrew see, for example, Th. Pahl, *Quellenstudien zu Luthers Psalmenübersetzung* (Weimar, 1931), esp. pp. 127–8, and J. Ficker, '*Hebräische Handpsalter Luthers*', *S.H.A.* (1919), 5. Abh., *passim*, esp. pp. 14–15, 21 ff.

² *W.A.* vol. II, p. 33 (26–35). A paper of mine on the philological details of Luther's interpretation is being published in *The Journal of Theological Studies*, 1955.

³ '...ut sub illo sitis: a quo est intellectus et eruditio'; this is taken from Augustine, *Enarratio in Psalmos* (P.L. vol. XXXVI, col. 72). Luther's dependence on Augustine in Ps. ii. 10 is not mentioned by A. Hamel, *op. cit.* p. 226.

THE INSPIRATIONAL VIEW: LUTHER

the connexion between the literal and theological interpretations, and used the grammatical explanation as a basis for the theological exegesis. In this process he changed the meaning of Reuchlin's words until they were in full harmony with his theological view. He who tries to disentangle the different trends of Luther's thought in this exegesis of Psalm ii. 10, may well ponder the affinity of Luther's inspiration, Lefèvre's method of interpretation of the Psalms, and the thought of Augustine.

If the word 'to be wise' has the meaning 'to be made wise by God', the question arises how man can attain wisdom. Is it possible to understand truth through human rational effort or only through God's grace? It is important to discuss the answer Luther gives to this question in the *Lecture on the Psalms*, since it was delivered before the publication of the first edition of Erasmus's New Testament in 1516.

Luther makes his position perfectly clear when he applies his explanation of 'be wise' to the religious sphere:

To be made wise [intellectificatio] is not to be understood according to human wisdom but according to the Spirit and 'the mind of Christ' [1 Cor. ii. 16] about which the Apostle speaks beautifully in 1 Corinthians, chapter 2, saying that only those who are spiritual and believers have this knowledge. In short: celestial, eternal and spiritual, and invisible things can be known through faith alone, that is 'Things which eye saw not, and ear heard not, And *which* entered not into the heart of man' [1 Cor. ii. 9], which no philosopher and none of the men, 'which none of the rulers of this world knoweth' [1 Cor. ii. 8].[1]

Luther discovers in the Psalms the same thought that he finds in the Epistles of St Paul: the rejection of human wisdom; man must be spiritual and he must believe before he can understand Holy Writ; his whole disposition must be in conformity with Scripture if he wishes to grasp its real truth. He asserts:

Nobody is able to speak worthily or to hear any part of Scripture if his disposition of mind is not in conformity therewith so that he feels inside what he hears or speaks outside and says: 'Eia vere sic est'.[2]

---

[1] *W.A.* vol. III, p. 171 (31 ff.). The translation is borrowed from the R.V.
[2] *W.A.* vol. III, p. 549 (33–5); cf. vol. III, p. 191 (17 f.).

THE INSPIRATIONAL VIEW: LUTHER

Thus the experience of the reader must be in some way similar to that expressed by the words in the Bible. When this is not the case, he is unable to grasp its full significance.[1] Therefore Luther sometimes admits in his lectures that he is unable to speak 'worthily' about a Psalm or some text because he happens to lack the predisposition required.[2] The prerequisite of any understanding is to have a knowledge of Christ (scientia et notitia Christi) since He, 'the sun and the truth in Scripture',[3] is spoken of everywhere in the Bible and especially in the Psalms.[4] The deeper the understanding of Christ, the better will be the understanding of the Psalms and of the Bible. Therefore in the last resort it is God's inspiration or His grace that guides man to a true interpretation of Holy Writ. These thoughts lead back to the starting point: only a prophet is able to understand prophecies; or in other words, a man wishing to understand 'spiritual things' must be imbued with this very spirit.[5]

In this way theology rises high above human philosophy where only one of the human faculties, the intellect, is required. For the understanding of the Bible more is required, namely 'the whole heart' (Psalm cxviii. (cxix) 10) with its capacity for feeling.[6] Without this Holy Scripture cannot be understood.

Many speculate wisely [Luther writes] but nobody is wise in Scripture and understands it if he does not fear the Lord. And he who fears more, understands more. For 'the fear of the Lord is the beginning of wisdom'.[7]

Thus there is a contrast between human speculation and human wisdom on the one hand, and celestial wisdom on the other.

---

[1] *W.A.* vol. III, p. 549 (32–3).
[2] *W.A.* vol. III, p. 643 (32–3). For further examples see K. Holl, *op. cit.* p. 549 n. 3.
[3] *W.A.* vol. III, p. 620 (2–7).
[4] *W.A.* vol. III, p. 46 (17–23); vol. IV, p. 379 (35–6).
[5] *W.A.* vol. IV, p. 305 (10–12).
[6] *W.A.* vol. IV, p. 282 (8–9): '*In toto corde meo* non dimidio, ut qui solum velut philosophi intellectu exquirunt sine affectu, exquisivi.' Cf. *W.A.* vol. IV, p. 7 (22): 'Cor... voluntatem et affectum significat.'
[7] *W.A.* vol. IV, p. 519 (1–5), on Psalms cx. (cxi) 10. For similar statements in the *Lecture on the Psalms* see H. Böhmer, *op. cit.* p. 48 n. 2.

THE INSPIRATIONAL VIEW: LUTHER

Nobody who uses his worldly knowledge is able to understand the truth of Holy Writ. Luther asserts that it is not in human power and not within the compass of human intellect to understand God's word. Therefore one should mistrust the intellect when expounding the Bible. God will grant us true understanding only through our humility and prayer. God turns away from those who, proud, obstinate and puffed up, rely on their own perceptions.[1] It must be clearly recognized that for Luther there is no bridge leading from human reasoning to the understanding of Holy Writ. Those who do not see the fundamental difference between human and divine are bound to distort the truth of God's word, which they submit to their own subjective reasoning or to their own personal experience.

Yet Luther himself says that for the understanding of the Bible man's disposition of mind must be in conformity with Scripture. Does not this view open the way to that subjective exegesis of the Bible which, as mentioned above, was rejected by men like Sebastian Brant at the end of the fifteenth century? If human reasoning cannot penetrate God's truth, and if therefore philosophy is to be excluded from Bible exposition, the traditional method of Bible interpretation as practised by the Schoolmen loses its validity. There seems to be no way of eliminating subjectivity except by an appeal to an authority able to decide whether an interpretation is true or not. In the years of these *Lectures* Luther still found this authority in the Church. Two passages may be quoted here which show not only that Luther was fully aware of the gulf between philosophy and theology but also that he was a loyal son of the Church. In his *Lecture on the Psalms* Luther wrote:

One should not do with Holy Scripture as is done with Aristotle where the wise may contradict the wise. For in philosophy the quality of the master decides the doctrine: a profane master and a profane doctrine. But with Scripture it is different: a holy master and a holy doctrine. Thus if anywhere by anybody a meaning is made known which is not

[1] *W.A.* vol. III, pp. 516 (39)–517 (39); cf. vol. IV, p. 319 (24–6).

THE INSPIRATIONAL VIEW: LUTHER

opposed to the articles of faith (regulis fidei), this meaning ought to be opposed by nobody, and nobody ought to prefer his own meaning, even if it be more evident and more in harmony with the letter.[1]

And in the same *Lecture* he said:

He who is not content with the Gospel and with God's word which was once preached throughout the world and which has been confirmed by so many martyrdoms of the saints, and he who dares to erect another doctrine and another wisdom, refusing to listen to the books of the Apostles and the elders of the Church, and he who is puffed up by his carnal understanding: this man rebels against God with horrible temerity. For in this way heretics deny Christ (that is the truth in the Church) and hence .... They seek to gain another truth which in their judgement is better and saner, and they seek to gain it in such a way as to wish scornfully that the truth which is in the Church, neither exists nor appears to exist.[2]

The submission to the judgement of the Church keeps the new, original reasoning in check,[3] and the danger of purely subjective interpretation is averted. Luther's opinion that man can find the true meaning of Holy Writ by God's grace only, and not by human endeavour, is in agreement with his doctrine of justification by faith which eventually led to a breach with the Church. This doctrine says that man is not justified by works but by faith only. In its application to the interpretation of the Bible it implies that studies do not aid man towards the understanding of God's word. Man must become humble; that means, according to Luther, that he must be led to a recognition of God and to humility. It is thus God who leads to a true perception of the Bible. Luther's ideas about the interpretation of the Bible are

---

[1] *W.A.* vol. III, p. 517 (33–9).

[2] *W.A.* vol. III, pp. 577 (39)–578 (7): 'Quicunque non est contentus de evangelio et verbo dei semel per mundum praedicatum et tot martyriis sanctorum confirmatum, et audet aliam doctrinam et sapientiam erigere, respuens audire libros Apostolorum et maiorum Ecclesiae, inflatus sensu carnis suae; hic horribili temeritate tentat Deum. Sic enim Haeretici Christum negant, id est veritatem in Ecclesia, et per consequens *Christus quoque: aliam quaerunt veritatem et suo iudicio meliorem et saniorem, ac ita quaerunt, ut eam, quae in Ecclesia est, fastidientes cupiant nihil esse et videri.' (*The reading 'Christus' is obviously corrupt.)

[3] For further examples see H. Preuss, *op. cit.* p. 15 and notes 10–16.

THE INSPIRATIONAL VIEW: LUTHER

thus a direct outcome of his inspiration. Hence it may be concluded that he will cling throughout his life to the essential parts of his early theory of Bible exegesis.

The subjectivism of this method was held in check only as long as Luther respected the authority of the Church. When he broke with Rome, there was no authority left to decide the validity of personal reasoning in matters of interpretation.[1] As he believed that he had learned the true meaning of the Holy Writ from God, he had to condemn as heretics (*'Rotten'*) all those whose views disagreed with his in any point, even if these heretics also claimed to be inspired by God. Thus one of the difficulties the Reformer had to face can be traced back to the exegetic principles of Luther's early life.

That Luther's belief in God's grace was the predominant motive in his attitude to contemporary trends of thought will be confirmed in the discussion of his view of scholasticism and humanism.

His criticism of the Schoolmen may, at the beginning, have been influenced by the humanistic attacks which were directed against their philosophical method. Luther's mistrust of philosophy naturally led to censure of the Schoolmen, a censure in which he and the humanists could join arms. His reasoning, however, was entirely his own. The weapons he used were not philological but theological in nature.

He pointed out that he did not approve of a satire like the *Epistolae Obscurorum Virorum* or of Erasmus's wit which forced people to laugh and to jest at the afflictions of Christ's Church while every Christian should bewail these conditions with loud

---

[1] Cf. *Epistola Lutheriana ad Leonem Decimum summum pontificem. Tractatus de libertate christiana*, of 1520 (*W.A.* vol. VII, p. 47 (28–30)): 'Deinde leges interpretandi verbi dei non patior, cum oporteat verbum dei esse non alligatum, quod libertatem docet omnium aliorum.' This passage reads in Luther's German version (*W.A.* vol. VII, p. 9 (29–31)): 'da tzu mag ich nit leyden regel oder masse, die schrifft auszzulegen. Die weyl das wort gottis, das alle freyheyt leret, nit soll noch musz gefangen seyn.' Cf. A. Harnack, *op. cit.* vol. III, pp. 869 ff. (cf. pp. 878 ff.); E. Troeltsch, *Die Soziallehren der christlichen Kirchen und Gruppen* (Gesammelte Schriften, I) (Tübingen, 1912), pp. 462–3, 808 f. (in the English translation by O. Wyon (London, 1931), vol. II, pp. 485–7, 698 f.).

lamentations.¹ The Schoolmen and Aristotle, against whom Luther raged in great anger, are shown to have misled the Church. None of these philosophers, he asserted, was able 'to understand one chapter of the Gospel or of the Bible, not even one chapter of Aristotle'. The whole of Aristotle's *Ethics*, he pointed out, is very bad and moreover hostile to grace. Hence his proposition in the *Disputatio contra Scholasticam Theologiam* that only without Aristotle and without scholasticism could a man become a good theologian.² This proposition is the outcome of his view that philosophy, relying on human thought and not on God's grace, cannot lead to the understanding of God's word. 'I do not believe in a reformation of the Church without completely rooting out canon law, decretals, scholastic theology, philosophy, logic, as practised now, and without the institution of other studies', he wrote in May 1518.³ But a year earlier, in May 1517, he had pointed out that at the University of Wittenberg biblical studies flourished, whereas scholasticism was neglected and Aristotle went down 'to a doom everlasting'.⁴ In his opinion he had the right freely to criticize the views of all the Schoolmen if their sayings were not based on the biblical text.⁵

The humanists, especially Erasmus, and Luther repudiated the Schoolmen for different reasons. Erasmus rejected them because, owing to their ignorance of the original languages, they were unable to understand the Bible. Luther, however, was opposed to them because they transferred human doctrine and human thought into the realm of theology where only God's grace, and

[1] L.E. vol. I, no. 24 (11–13), of 5 October 1516; no. 25 (5–7) of [?] the same date; no. 50 (3–7) [beginning of November 1517].

[2] *Lecture on the Romans*, ed. L. Ficker, S. pp. 108 (21)–110 (23), of 1515–16; L.E. vol. I, no. 65 (36–42), of 24 March 1518; *Disputatio contra Scholasticam Theologiam*, written shortly before 4 September 1517 (W.A. vol. I, p. 226 (41–53)).

[3] L.E. vol. I, no. 74 (33–6), of 9 May 1518: '...ego simpliciter credo, quod impossibile sit ecclesiam reformari, nisi funditus canones, decretales, scholastica theologia, philosophia, logica ut nunc habentur, eradicentur et alia studia instituantur.'

[4] L.E. vol. I no. 41 (8–13), of 18 May 1517; cf. nos. 42 (26–9), 65 (38–42), of 24 March 1518; 74 (30–40), of 9 May 1518; 117, of 9 December 1518; 161 (7–14), of 13 March 1519; 174 (31–4), of 15 December 1519, etc.

[5] *Resolutiones Disputationum de Indulgentiarum Virtute*, of 1518 (W.A. vol. I, p. 530 (4–12)).

THE INSPIRATIONAL VIEW: LUTHER

not human logic, can lead to an understanding of Scripture. Thus, in 1519, looking back at the time when he was learning with the Schoolmen, he wrote: 'I had lost Christ there, now I have found him in Paul.' This statement should not be used for conclusions about Luther's view of scholasticism whilst he was studying its philosophy. It reveals, however, the similarity of his thought in 1519 and 1542/3, when he remarked that before his inspiration he had no real understanding of the Bible. To this may be added the words he wrote in 1545 on his inspiration: '...this passage of Paul was truly the gate of paradise', to prove that in 1519 it was his inspiration which determined his negative attitude towards scholasticism.[1]

After the break with Rome Luther expanded the field of criticism to include the Fathers of the Church who had used the allegorical method of exegesis. He mentioned the names of Gregory, Jerome, Cyprianus, Augustine, and Origen. These Fathers, he is reported to have said in a sermon of 1524, followed their own subjective views instead of St Paul to whom the Holy Spirit is the active force in Scripture. Nobody, he said in the same sermon, should interpret God's word unless he be inspired.[2] These words, derived from his own religious experience, attack the orthodox view and the traditional method of exegesis, for they imply that the Fathers of the Church were not inspired whenever they interpreted the Bible allegorically. Their authority in this respect is, according to Luther, worth nothing, since they follow their own subjective views. It has been mentioned above that those who followed the orthodox view reproached their opponents for introducing their own personal opinions and thoughts into the exegesis of Holy

---

[1] For Luther's attitude towards the Schoolmen see O. Scheel, *Martin Luther* (Tübingen, 1930), vol. II, pp. 251 ff.; *Resolutiones Lutherianae super Propositionibus suis Lipsiae Disputatis*, of 1519 (*W.A.* vol. II, p. 414 (28) = Scheel, p. 14 (36); *T.R.* vol. V, no. 5553, of winter 1542/3 = Scheel, p. 173 (3–18); no. 5247, dated between 2 and 17 September 1540; cf. no. 5693 of ? 1542 = Scheel, pp. 162 (12–20), 163 (12–16)).

[2] *Predigten über das zweite Buch Mose* (*W.A.* vol. XVI, pp. 67 (12)–74 (36), esp. p. 68 (22 ff.) = Scheel, p. 35 (35–7)): 'Die ursach ist diese, das sie alle irem dünckel, kopff und meinung, wie sie es recht angesehen, und nicht S. Paulo gefolget haben, der da wil den heiligen Geist drinnen lassen handeln....'

Writ. Luther levelled this censure against the patristic allegorical interpretation and thus against the traditional view.

Luther's attitude towards humanism is more complex. The influence of Lefèvre's writings on him was indicated in the earlier parts of this chapter. He had become suspicious of the Latin translation of the Vulgate, and in his *Lecture on the Romans* of 1515–16 he took his references to the Greek original from Lefèvre's *Epistolae Divi Pauli*.[1] But this work contained only selected passages of the Greek. When Erasmus's New Testament was published in 1516, Luther at once made use of this Greek edition, and his first reference is found in Chapter IX of his *Lecture on the Romans*.[2] Erasmus's influence can be discerned in all the lectures which Luther gave between 1516 and his translation of the New Testament in 1522.[3] All these lectures may be dealt with together, although Luther doubtless made some progress in his method of interpretation and in his knowledge of Greek during this period of his life. He could therefore criticize Erasmus's philology better in 1521 than in 1516 when he lacked the necessary linguistic knowledge.

But for the first time Luther had a complete Greek text, and with the help of Erasmus's notes he could compare the text of the Vulgate with the original. It is therefore not surprising to find that the number of his references to the Greek wording increases greatly from 1516 onwards.[4] Erasmus's notes contain many re-

[1] E.g. Rom. viii. 33 (Ficker, *M.Gl.* p. 79 (21–4)), Lefèvre, *Epistolae divi Pauli apostoli* (Paris, 1512), fol. lxix^v (last lines).

[2] *Luthers Vorlesung über den Römerbrief 1515/16*, ed. J. Ficker, in *Anfänge reformatorischer Bibelauslegung*, vol. I (4th ed. Leipzig, 1930), pp. xlvi, liv.

[3] These lectures are: (*a*) *Luthers Vorlesungen über den Galaterbrief 1516/17*, ed. H. von Schubert, *A.H.A.* 5 Abh. (1918); cf. for emendations: H. von Schubert-K. Meissinger, 'Zu Luthers Vorlesungstätigkeit', *S.H.A.* (1920), 9 Abh., pp. 36–47; cf. J. Ficker, 'Zu Luthers Vorlesungen über den Galaterbrief 1516/17', *Theologische Studien und Kritiken*, 98/99 (1926), pp. 1–17. (*b*) *Luthers Vorlesung über den Hebräerbrief 1517/18*, ed. J. Ficker, in *Anfänge Reformatorischer Bibelauslegung*, vol. II (Leipzig, 1929); another edition (which I have not seen) is by E. Hirsch-H. Rückert, in *Arbeiten zur Kirchengeschichte*, vol. XIII (Berlin, 1929); cf. J. Ficker, 'Luther 1517', *Schriften des Vereins für Reformationsgeschichte*, vol. XXXVI (1918), pp. 14ff. (*c*) *In Epistolam Pauli ad Galatas...commentarius, 1519* (*W.A.* vol. II, pp. 436–618). (*d*) *Operationes in Psalmos 1519–21* (*W.A.* vol. V, pp. 1–676).

[4] E.g. Rom. ix. 15 (Ficker, *M.Gl.* p. 85 (24–5)); cf. Gal. ii. 2 (Schubert, p. 8 (24–7)).

marks on the meanings of words and many grammatical observations to which no objection could be made. The study of this detailed linguistic commentary had the effect that Luther paid more attention to the individual word in its grammatical environment. He often preferred the original Greek to the text of the Latin Vulgate. This implies, even though he does not expressly state it, that he does not consider the Vulgate as an authority which can be used for the interpretation of the Bible. But he is unwilling to say that St Jerome had mistranslated the Greek text. In the *Lecture on the Galatians* of 1516/17 Luther mentions the fact that Jerome censured earlier translators. As the text of the Vulgate is at some places identical with that of the earlier versions Jerome cannot, he maintained, have been the translator. This statement,[1] based on Lefèvre and Erasmus, is very important, for the authority of St Jerome is thereby saved. Neither Luther nor Lefèvre wished to correct Jerome's Bible, and Erasmus had, in a similar vein, asserted that Valla in his *Adnotationes* had mostly corrected those words which Jerome had taken over from earlier translators.[2]

Reference to the original languages of course influenced the interpretation; instead of using the literal meaning of the Latin wording, Luther had to explain the Greek text. Unlike the 'prophetic' sense which in his first *Lecture on the Psalms* was thought to coincide with the literal sense, Luther turned his attention to the real meaning of the Greek words. In this he naturally followed Erasmus's interpretation. Both his interlinear and marginal glosses contain many words taken from Erasmus's text or his notes. Erasmus thus provided the philological interpretation of the biblical text. Luther used the Vulgate as his text in all his lectures delivered before he began the Bible translation and he drew the students' attention to changes of wording or meaning

---

[1] Gal. v. 9 (Schubert, *M.Gl.* p. 24 (26–30)); cf. Gal. v. 4 (Schubert, *M.Gl.* p. 23 (26) and S., p. 61 (27f.)); Lefèvre, *Epistolae divi Pauli ad loc.* fol. cxxvi^r, no. 26; Erasmus, *N.T.1, ad loc.* p. 518, last two lines. Schubert, in his edition of Gal., refers to Luther's dependence on Er. *N.T.1*, p. 518.

[2] Lefèvre, *Epistolae divi Pauli* (Apologia, fol. a ii^v); *E.E.* vol. I, no. 182 (149–51).

THE INSPIRATIONAL VIEW: LUTHER

which resulted from the study of the Greek original. As an example I would mention Luther's interpretation of Romans xiv. 14.

The Latin text of the Vulgate is 'scio et confido', Erasmus's translation of 1516 is 'novi siquidem et persuasum habeo'. In his Annotations he explains the Greek text (οἶδα καὶ πέπεισμαι) with the words: 'scio et certus sum, sive persuasum habeo' where 'certus sum' is a paraphrase of the literal translation 'persuasum habeo'. In his interlinear gloss Luther interprets the words of the Vulgate as 'certus sum' which is taken from Erasmus, without, however, mentioning Erasmus's name. In the marginal gloss Luther writes: 'Hoc teutonica dicitur: Ich weysz vnd bynnsz gewisz.'[1] This is an exact translation of Erasmus's note, and corresponds to the text of Tyndale's version: 'For I knowe and am full certified.' In the translation of this verse we have an example of how Erasmus's interpretation influenced the translations of the Bible into the vernacular.[2] The treatment of the Bible text by Luther shows that, although objecting on principle to Erasmus's work, Luther often followed him in details of exegesis.[3] Other examples of Luther's dependence on Erasmus are: Luther repeats Erasmus's corrections of the Vulgate when they are based on the Greek text;[4] he borrows grammatical explanations, as, for example, the construction of a sentence;[5] he even mentions the

---

[1] To Rom. xiv. 14 (Ficker, *I.Gl.* p. 124 (4) and *M.Gl.* p. 124 (19)).

[2] The English translations of this verse are: Wyclif: 'I woot and triste'; Tyndale (1534): 'For I knowe and am full certified'; Cranmer (1539) is identical with Tyndale; Geneva (1557): identical with Tyndale but omitting 'for'; Rheims (1582): 'I know and am persuaded'; A.V. (1611) and R.V. (1881) are identical with Rheims.

[3] A few examples of Erasmus's influence on Luther's exegesis are: Rom. xv. 26 (Ficker, *I.Gl.* p. 135 (15)) which, as the Greek word proves, is taken from Erasmus and not from Lefèvre. (Ficker refers to both these humanists as a possible source.) Rom. xiv. 16 (Ficker, *I.Gl.* p. 125 (1–3)); Rom. xiv. 22 (Ficker, *I.Gl.* p. 126 (7–8), *M.Gl.* p. 126 (23–6)) where Luther mentions Erasmus as being his source for a reference to Ambrose; Gal. i. 6 (Schubert, *I.Gl.* p. 4 (5–6), *M.Gl.* p. 4, (17–19, 24)); Gal. i. 6, Revision of 1519 (*W.A.* vol. II, p. 460); Hebr. viii. 13, ix. 1 (Ficker, *I.Gl.* p. 37 (3–10)).

[4] E.g. Hebr. v. 11 (Ficker, *I.Gl.* p. 23 (12)) where Luther agrees with Faber's contents though not with the wording which is Erasmus's.

[5] Rom. xiv. 19 (Ficker, *I.Gl.* p. 125 (14) and *M.Gl.* p. 125 (26–7)); Gal. ii. 10 (Schubert, *M.Gl.* p. 9 (31)); Gal. ii. 10, Revision of 1519 (*W.A.* vol. II, p. 483 (14)); Gal. ii. 16 and ii. 20 (Schubert, *I.Gl.* pp. 11 (8), 12 (6) and *M.Gl.* p. 12 (23)).

use of rhetorical forms such as allegories or paradigmas.[1] The paradigma is used, Luther says, to make the meaning clear to those who are not learned. In these explanations, which may occur in places where they are not found in Erasmus's notes, Luther is influenced by Erasmus in two ways: in the consideration of figures of speech, and in the attention paid to a given situation which makes the usage of special words necessary. Luther even makes references to stylistic phrases which are described as not being characteristic of St Paul. These observations are probably imitations of Erasmus's method.[2]

Luther follows Erasmus in philological details. Indeed there is hardly a page in Luther's lectures where Erasmus's influence is not felt. But it would be wrong to assume that Luther neglected other sources. He used the ordinary gloss and the Fathers of the Church; Reuchlin, Lefèvre and Erasmus showed him the value not only of the original texts for the interpretation of the Bible but also of those expositions whose authors were versed in Hebrew and Greek. Therefore Lyra, for instance, is no longer rejected because of his preference for the Hebrew text, as he was in Luther's *Lecture on the Psalms* of 1513–15.[3] In the *Lecture on the Galatians* Luther even expressly states that he prefers Jerome's exegesis of this letter to that of Augustine.[4]

[1] Gal. iv. 24 (Schubert, *I.Gl.* p. 21 (13–14)); Hebr. vi. 7 (Ficker, *I.Gl.* p. 25 (6) and *M.Gl.* p. 25 (19–22)). This is influenced by, but not borrowed from, Erasmus. For later times cf., for example, Luther's words (in *Auff das ubirchristlich...Buch Bocks Emszers... Antwortt*) of 1521: 'Solch blumen wortt leret man die knaben ynn den schulen und heyssen auff kriechsz Schemata, auff latinisch figure, darumb das man damit die rede vorkleydett unnd schmuckt.... Der selben blumen ist die schrifft voll, sonderlich ynn den propheten...' (*W.A.* vol. VII, p. 651 (31–4)).

[2] E.g. Hebr. ii. 10 (Ficker, *I.Gl.* p. 9 (8) and *M.Gl.* p. 9 (14 ff.)); cf. Rom. xiv. 19 (Ficker, *M.Gl.* p. 125 (23–5)).

[3] Cf. above, p. 176. For Luther's sources see Ficker's editions of Romans, pp. liii–lv, of Hebrews, pp. xxxiii–xxxiv. It is, however, very doubtful if Luther studied rabbinical literature as Ficker's wording (Hebrews, p. xxxiii) may imply. It seems more probable that Luther knew it second-hand only; cf. Ficker's review on M. Freier, 'Luthers Busspsalmen und Psalter', *Beiträge zur Wissenschaft vom Alten Testament*, part 24 (1918). The review is in *Theologische Literaturzeitung*, vol. XLIV (1919), p. 59. It is difficult to prove the influence of the mysticism of Tauler and *Theologia Teutsch* on Luther. This is the view held by Schubert in his edition of Galatians, p. xiii, and of O. Scheel, 'Taulers Mystik und Luthers reformatorische Entdeckung', *Festgabe für D. Dr. Julius Kaftan zu seinem 70. Geburtstage* (Tübingen, 1920), pp. 298–318. But Ficker in his edition of Romans believes that this influence is very important (pp. lxxxi ff.).

[4] Gal. iv. *initio*; Schubert, *M.Gl.* p. 18 (23).

THE INSPIRATIONAL VIEW: LUTHER

When considering these facts, it is possible to see humanistic influence on Luther in its proper perspective. Luther's suspicions about the accuracy of the Vulgate were confirmed and strengthened by Erasmus's criticism. He paid closer attention to the meaning of the words. In short, he followed the technique of humanistic scholarship.

Yet the differences are clear from the very beginning. Erasmus was satisfied to interpret Holy Writ from the standpoint of the grammarian and refused to be embroiled in detailed theological discussion. He explained the meaning of a word and left it to theologians to determine theological definitions. Luther used the philological method as worked out by Erasmus but the main part of his work is the theological interpretation. Luther's notes begin at the point where Erasmus's end. Erasmus explained the word in its context, Luther went on to discuss its religious significance.[1]

An example may be given here to illustrate Luther's use of, and his reaction to, Erasmus's linguistic commentaries. In 1518 Luther wrote about the meaning of the word 'penitence'. He explained how during his life the meaning of this word underwent several changes for him. First Staupitz explained it to him in a way different from the traditional interpretation, then he learned from 'very erudite men' who knew Hebrew and Greek that the real meaning of this term is 'change of mind and the understanding of one's own sin after having suffered harm and having recognized the mistake'. This definition is taken from Erasmus, who had divided the Greek word into its individual parts and found that the recognition of the evil committed leads to the feeling of sin and thus to a change of mind.[2] It is interesting

---

[1] See, for example, Gal. v. 10: Erasmus's *N.T.1*, p. 519: the words 'qui vos conturbant' (οἱ ἀναστατοῦντες) are explained 'id est qui vos a statu dimovent'. Luther (Schubert, *I.Gl.* p. 24 (15)) interprets 'conturbant' as 'de recta fidei doctrina deiecit' and in the revision of 1519 he enlarges upon it (*W.A.* vol. II, p. 571 (18–21)): '"Conturbant" id est, de vera fide doctrinis suis deiecit ac deturbat a statu, in quo stabatis (cf. Erasmus). Sed nunquid excusabit illum pius zelus et bona, ut dicunt, intentio? aut ignorantia? aut quod Apostolorum discipulus est et magnus? Non, inquit....'

[2] Luther, *Resolutiones disputationum de indulgentiarum virtute* of 1518 (*W.A.* vol. I, p. 525). Erasmus, *N.T.1*, to Rom. ii. 4, p. 424 and to Matth. iii. 2, *N.T.1*, p. 241. A detailed study on this subject will be published by me in *The Journal of Theological Studies*, 1955.

THE INSPIRATIONAL VIEW: LUTHER

to note that Luther explained this definition with words which, in their ambiguity, may easily change Erasmus's basic conception, for he continues that this change of mind 'cannot take place without a change of sentiment and love'. 'All this', he goes on, 'corresponds closely to the theology of St Paul; thus, in my opinion at least, there is scarcely anything that can illustrate Paul more appropriately.' These words reveal the working of Luther's mind: he took Erasmus's grammatical explanation but assimilated it to demonstrate his own religious ideas. From here he proceeded to a new stage, in which he again begins with an analysis of the Greek word. But this is a new analysis, a 'forced' one as he called it, which enabled him to explain the Greek word as 'a transmutation of mind and sentiment'. This, in Luther's view, 'seems to manifest God's grace'. Thus a transmutation, necessary for penitence, is connected with God's grace, a thought which is parallel with Luther's tenet of 'justification by faith'.[1] In this way Luther has reached a new climax. The popular belief in the meaning of 'penitence' has been left far behind. Staupitz had helped him, then he learned Erasmus's explanation of this word, examined it and replaced it by a theological definition which although based on a grammatical *tour de force* was primarily theological. It is therefore only to be expected that what he called '*a transmutation of mind and sentiment*' was for him 'the most rigid translation' of *penitence*. This he paraphrased: 'put on a different mind and sense, return to your mind (this is Erasmus's rendering), effect a change of mind and a change of spirit'. He terminates this enumeration with the quotation of Romans xii. 2: 'Be ye transformed by the renewing of your mind.'[2]

The principal difference between the exegesis of Erasmus and that of Luther may be stated as a conclusion. Erasmus analysed the Greek word and explained it as relating to a rational process: something wrong had been done. This is recognized and thus a change of mind achieved. Luther accepted Erasmus's philological explanation because 'all this closely corresponds to the theology

[1] *W.A.* vol. I, p. 526 (1-4).　　[2] *W.A.* vol. I, p. 530 (19-24).

of Paul'. This sentence would, I think, not have been understood by Erasmus, who would have considered it as a truism, since for him the real meaning of the text must be based on philology. For Luther the philological meaning is known and the analysis of the Greek word is used to support the theologian's view. The methods and beliefs of Erasmus and Luther were thus utterly opposed to one another. Erasmus based his interpretation of the text on grammar and knowledge of philology; for Luther it was theology that governed grammar. Therefore the proof of his own word explanation lay in its agreement with his theological principles even though this might mean a forced interpretation.[1] There was thus, in Luther's opinion, a connexion between the philological explanation and the theological contents. The theological meaning had to be known before grammatical considerations could be applied to the text. But there was no way from grammatical analysis to theological understanding.

If this thought is pursued to its logical conclusion, it leads to the denial of all learning. If the study of grammar and the learning of languages cannot lead to the true perception of God's word, why should human knowledge be acquired? It could be said that human wisdom is useless or even misleading. Luther pointed this out in 1518, when he maintained the paradox that the wiser man is in learning, the less he understands God's word.[2] This is an attitude similar to that of St Augustine quoted above: 'The weaker man is,...the more learned he wishes to appear, learned not in the knowledge of those matters which edify man, but in the knowledge of signs only, which most easily inflame the mind.'

There were indeed followers of Luther who objected to all learning and put their trust in illumination only. They thought,

[1] J. Soury ('Luther, Histoire de sa préparation exégétique', *Revue des deux Mondes*, 15 October 1871) was the first to write a paper on Luther's exegesis. He writes (p. 929): '...la préparation exégétique de Luther prouve qu'il n'a été ni helléniste, ni hébraïsant, ni philologue au sens ordinaire de ces mots. Homme de foi et d'action, il n'a ni le goût de la science pure et désintéressée, ni le loisir....' Cf. K.A. Meissinger, *Luthers Exegese in der Frühzeit*, Dissertation (Giessen, 1910), p. 36: 'als Exeget ist Luther kein Gelehrter.'

[2] *Auslegung des 109. (110.) Psalms* of 1518 (*W.A.* vol. IX, p. 187 (5–12, 16–23)). This is a reprint of Luther's MS. The first printed edition is published *W.A.* vol. I, p. 696 (10 ff.).

THE INSPIRATIONAL VIEW: LUTHER

as Luther reported of Müntzer, that one should conceal oneself, offer one's empty heart to God and then truth would be revealed. Luther clearly saw that reliance on inspiration only may not merely distort God's word but actually replace it. He knew from his own experience that neglect of the biblical wording leads to flights of fancy. Only a return to the text of Scripture can prove that vain thoughts have no connexion with God's word but are based on purely human, and not divine, wisdom. Thus Luther connected the actual text of Holy Writ with the revelation essential for perceiving its real meaning. In this connexion he speaks of his own experience: only after study, he maintains, was he granted God's grace and he proclaims that only complete devotion to the study of God's word can open up its understanding.[1] These sentences of 1526 again confirm his report on his inspiration when in 1545 he wrote that 'for days and nights' he meditated about the connexion of the words in Romans i. 17 before their meaning was revealed to him. The order of events which led to his inspiration furnished Luther with a basis to fight against those of his followers who believed in illumination only: their revelations are not in agreement with the text of the Bible and therefore they can be proved to be wrong. His inspiration also provided him with a criterion for censuring the grammatical explanations of the humanists. They believed in the capability of man to understand God's word through learning, without considering that the sacredness of the letter was beyond the reach of human grasp. The gospel, Luther wrote in 1524, has been given to man through the Holy Ghost only, but His message was transmitted by means of language.[2] There is thus a close connexion between the content of Holy Writ and its actual wording in its original languages. It is necessary for the understanding of

---

[1] *W.A.* vol. XVI, p. 598 (12–33), *Predigten über das 2. Buch Mose* of 1526 = Scheel, p. 43 (1–15), where only part of this passage is printed.

[2] *W.A.* vol. XV, p. 37 (3–6). See the important sentences in *Vorrede zum 1. Bande der Wittenberger Ausgabe der deutschen Schriften* of 1539 (*W.A.* vol. L, p. 659 (22–35)). Cf., for example, R. Seeberg, *Lehrbuch der Dogmengeschichte* (4th ed. Leipzig, 1920–33), vol. IV, 1, pp. 28 ff.

THE INSPIRATIONAL VIEW: LUTHER

Scripture to study the exact meaning of the text in the original languages, for words are the vehicle of thought, they are like a shrine in which God's truth is found. These last words are, in all probability, borrowed from Luther's friend, the humanist Melanchthon.[1]

Indeed Luther's view on the importance of language contains many a humanistic feature. He is convinced that human civilization depends on the knowledge of languages. But he adds his own interpretation when he wishes to recognize God's providence in the propagation of Latin and Greek before the Christian era. God willed this in order to facilitate the dissemination of the Gospel. Similarly, Luther argues, it was God's wish that Greece should be overrun by the Turks, for the dispersion of the Greeks would be followed by a revival of Greek studies in Europe. Luther did not write a systematic treatise on this subject, but only mentioned these two items, without explaining why languages decayed after the time of the apostles.[2] He admits that the outcome of this decline in the knowledge of languages was the decay of Christianity which, at the end, was destroyed by papacy. The Fathers of the Church, he asserts, often erred and contradicted themselves because of their ignorance of languages. Thus it has come about that the Bible has been considered to be unintelligible. Nothing, however, is clearer than God's word if read in the original languages and without reference to the exegesis of the Fathers. Now that the biblical languages are learnt again, Holy Writ is seen in all its clarity, almost as at the time of the apostles. It can definitely be understood better than at the time of Jerome and Augustine. Languages should therefore be learnt, the Bible

---

[1] Luther's phrase 'Sie sind der schreyn, darynnen man dies kleinod tregt' (*W.A.* vol. xv, p. 38 (9)) is dependent on Melanchthon who in 1523 had written: '...vis verborum...in quibus tanquam in sacrario quodam divina mysteria recondita sunt' (*CR.* vol. II, col. 64). Cf. K. Hartfelder, 'Philipp Melanchthon as Praeceptor Germaniae', *loc. cit.* p. 205 (cf. p. 166).

[2] It is unnecessary to harmonize this argument of 1524 with later sayings of Luther's where he contrasts the great knowledge of languages and the 'blindness of the Church scarcely 300 years after Christ', as is reported in *T.R.* vol. IV, no. 5009 of 21 May to 11 June 1540.

THE INSPIRATIONAL VIEW: LUTHER

should be read in its original form, and the commentaries and glosses of old should be neglected. Only if this is done will the Bible be preserved in its purity.[1] Hence Luther could wish that his own writings would perish, for their mere existence involved the danger that they might be read instead of the Bible.[2] But he could also maintain that Melanchthon's exegesis excelled every interpretation of the Bible written during the last thousand years, and that his own translation was better than any other version made into Greek or Latin.[3]

Melanchthon indeed fulfilled the two requirements demanded by Luther: he had a great knowledge of languages, and he perceived that without the help of the Holy Ghost no understanding is possible. It may be useful to quote Melanchthon's words, in which this master of formulation brings out the twofold nature of Bible exegesis:

I have not the mistaken view that the holy can be penetrated through the industry of human talent. There is something in the holy that nobody can ever see unless it is shown to him by God: and Christ cannot be known to us without the Holy Ghost teaching us.... But apart from prophecy, the meaning of the words must be known in which, as in a shrine, the divine mysteries are hidden. For what is the use of reciting in a magic way words that have not been understood? Is it not like telling a story to a deaf person?[4]

Luther often made his position clear, never perhaps better than in 1539 when he pointed out that Holy Writ turns human wisdom to foolishness since it teaches eternal life, something the human mind cannot understand. Thus man cannot perceive the meaning of the Bible without the help of the Holy Ghost. Therefore he

---

[1] *An die Ratherren aller Städte deutsches Lands, dasz sie christliche Schulen aufrichten und halten sollen* (*W.A.* vol. xv, pp. 37–41). Cf. H. Lilje (*Luthers Geschichtsanschauung* (Berlin, 1932), p. 33), whose interpretation of this passage is not satisfactory.

[2] See Luther's words of 1539 (*W.A.* vol. L, pp. 657 (2–11), 658 (13–20)) and of 1545 (*W.A.* vol. LIV, p. 179 (13–14) = Scheel, p. 186 (24–5)); cf. his view of 1519/20 (*W.A.* vol. v, p. 20 (8–9)).

[3] *L.E.* vol. II, no. 316 (6–13), of 28 July [1520]; *T.R.* vol. I, no. 252 of 20 April to 16 May 1532; *L.E.* vol. II, no. 449 (52–4), of 13 January 1522. Cf. *T.R.* vol. v, no. 5324, of 19 October to 5 November 1540.

[4] *Encomium Eloquentiae* (*C.R.* vol. II, col. 64).

THE INSPIRATIONAL VIEW: LUTHER

should pray to God to receive illumination, guidance and understanding. But God will grant understanding only to him who perseveres in the study of the words themselves, and meditates upon the meaning with which they are informed by the Holy Ghost.

The interconnexion between God's guidance and human endeavour is thus clearly established: the theologian cannot interpret Scripture without learning and the scholar is incapable of writing a commentary on the Bible unless God's grace has illuminated him and revealed the true meaning of the text. The theologian who relies on translation, wrote Luther in 1524 after part of his own version had been published, may understand Christ and be a good preacher and teacher—but not an expositor of the Bible able to defend his exegesis against erroneous teaching. This task can be done by a prophet only.[1] It is important to notice that Luther calls an expositor of the Bible a prophet.[2] This proves that he was within the tradition of Philo and St Augustine, but his emphasis on the importance of the letter of Scripture is a significant development in this school of thought, as will presently be shown.

This analysis of Luther's inspirational principle enables us to understand his attitude towards Erasmus. Luther obviously respected the grammatical and linguistic achievement of Erasmus's edition of the New Testament. He was not in a position to criticize the philological notes when it was published in 1516 since his own learning was not sufficient. But the question derived from his inspiration was whether these grammatical explanations were the result of God's grace or of Erasmus's human reasoning.

In a letter of 1516 Luther pointed out that unlike Erasmus's view, he prefers Augustine to Jerome. He emphasized that Erasmus's explanation of the term 'righteousness' was wrong, for it has not the same meaning as in Aristotle's philosophy, where it is said that man will be righteous through acting righteously. On

---

[1] *An die Ratherren...* (*W.A.* vol. xv, p. 40 (14–26)).
[2] According to Luther a prophet either foretells the future or he expounds the Bible. See H. Preuss, *Martin Luther, der Prophet* (Gütersloh, 1933), pp. 84–94, esp. pp. 85, 91–3.

THE INSPIRATIONAL VIEW: LUTHER

the contrary, the biblical connotation teaches that man must first be righteous before he can act righteously. This theological interpretation of October 1516 is in the fullest possible harmony with his report on his inspiration dated 1545, an additional proof of the way in which Luther's thought was determined by his own experience. Luther seems to imply in this criticism that all Erasmus's learning could lead him only to an understanding of philosophical terminology but not to the realization that the same term used in theology has a meaning totally different from its usage in profane sciences. Erasmus's interpretation contains in Luther's opinion 'that literal, that means dead [mortua intelligentia], understanding with which Lyra's commentary and almost all the commentaries after Augustine are filled'. This judgement is also extended to Lefèvre who, however, is called 'spiritual and very sincere'.[1]

Erasmus's lack of spiritual perception is again emphasized in another letter which Luther wrote on 1 March 1517 when he was actually reading Erasmus's New Testament. This letter, marked 'confidential', defines the difference between Luther and Erasmus in one sentence: 'The judgement of him who attributes something to the free will of man is different from the judgement of him who knows nothing but grace.' These words indeed contain the basis of Luther's criticism of humanism, expressing clearly the principal difference between him and Erasmus which gave rise to the pamphlets in which they came out in open controversy in 1524, namely over the question of free will. If, Luther maintained, 'something is attributed to free will', man may by his own exertions be able to arrive at an understanding of God's word. If, however, 'there is nothing but grace', all these exertions are futile and condemned to failure unless God's grace is granted beforehand.[2] From this proposition the contents of Luther's letter readily follow: in Erasmus's exegesis of the Bible the human is

[1] L.E. vol. I, p. 27 (17–40), of 19 October 1516. Cf., for example, *Lecture on the Romans*, to i. 17 and 4. 7 (Ficker, S. pp. 14 (3) ff., 121 (10) ff.).
[2] For references to 'free will' see, for example, H. Preuss, *Die Entwicklung des Schriftprinzips bei Luther bis zur Leipziger Disputation* (Leipzig, 1901), p. 29.

predominant instead of the divine. From this it may be concluded that Erasmus has not received God's grace. Hence his knowledge of languages does not make him a wise Christian. An example is given to add more weight to this conclusion: Augustine, who did not know languages, is superior to Jerome the philologist. A pious wish results from this argument, namely 'The Lord will perhaps give Erasmus understanding in his own time', as Luther writes a few lines later in this letter.[1]

All this is in complete harmony with Luther's inspirational interpretation of the word righteousness. As man must be righteous before he can act righteously, so he must receive God's grace for the understanding of Holy Writ before his own knowledge can be of avail.

In this letter Luther again stresses the contrast between Augustine and Jerome, choosing Augustine as the better expositor of the two. But if the reader draws the conclusion that Luther despises all the learning of languages, he is mistaken, for this letter starts with the request that the recipient, Johannes Lang, should supervise the Greek studies of a fellow monk. A characteristic qualification is added however: these studies should be pursued in a Christian way (christianiter). Luther's condemnation of Erasmus implies that Erasmus's studies are not 'Christian'. Greek studies may thus be good or bad. The measure of their quality depends on whether the student has been converted by God to see His grace. If this is the case, the knowledge of languages seems to Luther to be useful. Therefore he advises a monk to learn Greek. It may be concluded that Erasmus's *New Testament* is not useful in its theological interpretation, although its grammatical explanations may be perfectly sound. The fact that, in Luther's view, Erasmus was unable to understand the spiritual meaning invalidates many of his grammatical explanations, for Erasmus could not understand the subject-matter, namely theology. A theologian, however, who has a true spiritual understanding may be able to make use of Erasmus as a grammarian so long as he critically avoids the dead

[1] *L.E.* vol. I, no. 35 (15–28), of 1 March 1517.

letter and penetrates into the spirit of Holy Writ. Luther can thus praise Erasmus the grammarian whilst at the same time he is bound to attack and to condemn Erasmus the theologian.

Erasmus with all his abilities is no danger to Luther's inner certainty. In 1522 Luther even writes that Erasmus's understanding of predestination is inferior to that of the Schoolmen. 'Mightier is truth than eloquence,' he continues, 'more important the spirit than genius, of greater weight faith than erudition.'[1]

This judgement on Erasmus is foreshadowed in the third letter written on 18 January 1518. Again this is a confidential letter, for, as Luther pointed out, he always praised the humanist in the presence of those who were opposed to *bonae literae*. He emphasized the difference between Augustine and Jerome. When speaking as a grammarian Jerome, who is highly praised by Erasmus, is by far the more capable of the two, but when speaking as a theologian, the judgement must be reversed. There is a great deal in Erasmus which seems unable to recognize Christ.[2]

But this letter does not only contain a criticism of Erasmus and Jerome; it also outlines the way in which comprehension of the Bible may be granted to man. Thinking of this part, Spalatin, to whom this letter is addressed, headed it 'Introduction to Theology'. This letter contains a reference to Luther's own inspiration as well as a refutation of the belief that man can 'penetrate into the meaning of the sacred books through study or mental ability'. Only through prayer and humility, Luther maintains, only through despair of study and only through trust in the flowing of the Spirit may God in His mercy grant man the true understanding of His words. 'For there is no teacher of His divine word but He Himself, the author of His word.' It is in this state of despair that one should begin one's studies: first the whole Bible should be read that the 'simple story' may be learned. For this Jerome's letters and commentaries are very useful. Then 'for the knowledge of Christ and of God's grace (that means for the

---

[1] *Iudicium D. M. Lutheri De Erasmo Roterodamo ad Amicum.*
[2] L.E. vol. I, no. 57 (10–29).

more secret understanding of the spirit) the blessed Augustine and Ambrose are more profitable'.[1]

Thus as early as 1518 Luther saw the polarity and interdependence of the two elements which alone could lead to an understanding of God's word. The despair in human capability, the complete humbleness required before God's grace may be granted to man, remind the reader of the 'empty heart' which Müntzer wished to offer to God for illumination. For Luther, however, this is the preparation for the study and understanding of Holy Writ. 'This understanding', he pointed out in his *Commentary on the Psalms* of 1519 to 1521, 'this understanding comes from faith.' Again in 1521 he proclaimed: 'If I know what I believe in, then I know what is written in Scripture, for Scripture contains nothing but Christ and Christian faith.'[2] The meaning of God's word is therefore simple,[3] and these sentences elucidate the reason why Luther could discard the traditional fourfold exegesis which he had rejected in his *Lecture on the Galatians* of 1516/17. It is interesting to note that Luther borrowed his argument against the fourfold meaning from Augustine.[4]

Faith is the most important requirement for any understanding of Scripture. This conception is in complete agreement with Augustine's thought. Luther's ideas are essentially in harmony with this Father of the Church, while his protagonist Erasmus followed Jerome. The controversy between the religious theory of Luther and the humanistic thought of Erasmus is a continuation of the earlier dispute between the two Fathers of the Church. But it should not be forgotten that these two discussions were separated by more than eleven hundred years and that for this

---

[1] L.E. vol. I, no. 57 (31–45). Cf. L.E. vol. I, no. 175 'Beilage', p. 397 (20–34). Cf. T.R. vol. I, no. 252, of 20 April to 16 May 1532.

[2] Interpretation of Psalm ii. 10 (*W.A.* vol. v, p. 69 (15 ff.), esp. (20–6). The second quotation is from *Der sehs und dreyssigist psalm David* (*W.A.* vol. VIII, p. 236 (18–22)).

[3] For references see K. Holl, 'Luthers Bedeutung für den Fortschritt der Auslegekunst' (1920), *loc. cit.* p. 551 and n. 2.

[4] Ed. Schubert, S. p. 60 (5–31). *In Epistolam Pauli ad Galatas...Commentarius, 1519* (*W.A.* vol. II, p. 551 (16)–552 (19)). In 1516/17 Luther refers to Augustine, *De Spiritu et Litera* (P.L. vol. XLIV, cols. 203, 215 f.), in 1519 to Augustine, *Contra duas epistolas Pelagianorum* (P.L. vol. XLIV, esp. cols. 593 ff.).

reason there are of necessity differences in the arguments. The choice in Luther's time was between the method of the Schoolmen and that of the humanists. He objected to scholasticism in all its forms and therefore took over as much of the humanistic method as was compatible with his religious belief. Thus in many points he could not agree with Augustine. Luther's assessment of the value of earlier Bible translations, for example, differed from Augustine's. Both started from the same belief: faith is the basis of the understanding of Holy Writ. But Luther is not sure that all the earlier translators had the true faith. He sometimes speaks of it as of a quality which leaves its imprint in the human heart, making it a 'Christian heart' which he contrasts with the 'unchristian, Turkish, pagan heart' that is excluded from the true understanding of God's word.[1] This point of view enables Luther to find his own criterion for reviewing earlier commentators and Bible translations.

His views on the Septuagint and the Vulgate are often found in remarks preserved in his *Table Talks*. Luther condemns the translators of the Septuagint in very harsh terms: they were 'not learned',[2] ignorant of Hebrew, their rendering is most inept, and they misunderstood the words and idioms of speech. Therefore St Jerome is preferred to them.[3] Finally, Luther reproaches the translators of the Septuagint for falsifying the text, since they were afraid of giving offence by attributing divinity to Christ.[4] It may seem strange that Luther should inveigh against the translators of the Septuagint, who were Jews, for their ignorance of Hebrew. In his view Hebrew had been corrupted during the Babylonian captivity to such a degree as to make its restoration impossible.[5] Thus after the Babylonian captivity the Bible could no longer be

[1] *Ein sendbrieff Von Dolmetzschen...*, of 1530 (W.A. vol. xxx, part 2, p. 640 (25–9)); *Vermanung zum Sacrament des leibs und bluts unsers herrn*, 1530 (W.A. vol. xxx, part 2, pp. 599 (26–35), 600 (21–6)).
[2] T.R. vol. IV, no. 5001, of 21 May to 11 June 1540.
[3] T.R. vol. I, no. 1040, between 1530 and 1535; T.R. vol. III, no. 3271 *a* and *b*, of 9 August 1532. The readings of some words differ in the various traditions of these utterings.
[4] T.R. vol. IV, no. 4896, of 6 to 16 May 1540.
[5] T.R. vol. I, no. 1040, between 1530 and 1535; T.R. vol. III, no. 3271 *a*, *b*, of 9 August 1532; T.R. vol. II, no. 2758 *a*, *b*, of 28 September to 23 November 1532.

understood by the Jews.[1] This alone was a valid reason for rejecting all the Jewish commentators of the Bible. In 1523 he pointed out that even the Jews did not know enough Hebrew and that therefore their interpretations could not be trusted. But in this connexion he made it clear that the knowledge of Hebrew in itself was not sufficient since only those who know Christ are able to perceive the full meaning of Holy Writ.[2] Thus the two trends of thought are joined: the Jews cannot understand Scripture because they are Jews and not Christians, and even their knowledge of Hebrew is so corrupt as to invalidate their commentaries. It follows that Luther did not believe in the inspiration of the Seventy Translators. Their falsification of the original and their lack of understanding made it necessary for him to translate from the original Hebrew and not from the Septuagint.

St Jerome's translation of the Bible into Latin, the Vulgate, is praised by Luther, though with the following reservation: Jerome did his rendering alone, without help. This is, in Luther's view, a formidable reproach, which is explained by the quotation of Matthew xviii. 20: 'For where two or three are gathered together in my name, there am I in the midst of them.'[3] Therefore Jerome's work, learned and admirable though it is, lacks the support of the Holy Ghost. This view expressed between 1530 and 1535 completely tallies with Luther's criticism of Jerome of 1518 when he asserted that Jerome is useful for learning the 'simple story' of the Bible but not for the 'knowledge of Christ and of God's grace'. From this point of view it can easily be understood that Luther criticized Jerome and Erasmus together, as for example in a *Table Talk* of 1540 where he is reported to have said:

Jerome is a prattler like Erasmus: he wished to speak grandiloquently but he did not succeed. He promises something to the reader but he

---

[1] *T.R.* vol. v, no. 5327, of 5 November 1540.

[2] *Preface to the Old Testament* of 1523 (*E.A.* vol. LXIII, pp. 23–4); *Von den letzten Worten Davids* of 1543 (*E.A.* vol. XXXVII, p. 3); *Sendbrief* (*W.A.* vol. XXX, part 2, p. 640 (25–32)).

[3] *T.R.* vol. I, no. 961, between 1530 and 1535. The two versions are preserved in Roerer's copy of Veit Dietrich and by Aurifaber. For a characterization of Aurifaber see E. Kroker's Preface to *T.R.* vol. III, pp. xxxiii–xxxvii.

fulfils nothing. But I am amazed that at that time, scarcely three hundred years after Christ, so great a blindness was found in the Church together with so great a knowledge of languages.[1]

Yet Luther believed that Jerome's translation was very good, even the best version in existence. From the very beginning he wished his Bible to be compared with the Vulgate. When rendering the New Testament he wrote to Melanchthon that his work was a burden beyond his strength and that he would be unable even to start translating the Old Testament. But he expressed his hope that his German Bible 'may become a translation worthy to be read by Christians'. 'I hope', he continued, 'we shall furnish our Germany with a better version than that of the Latins.'[2] In 1523, in one of the prefaces to his version of the Old Testament, he was convinced 'that this German Bible is clearer and more accurate than the Vulgate at many places'.[3] In 1540 in one of his Table Talks he was sure that his translation was better than 'all versions Greek and Latin' and that it contained better interpretations than all commentaries.[4]

There is one reason for Luther's criticism of earlier Bible translations and for the belief that his own version was better than all the others. It is the belief that his understanding of the biblical text is greater than that of all his predecessors. He was convinced that he had received from God what Jerome and Erasmus had not, 'the knowledge of Christ and of God's grace'. This does not lead to the assumption that he understood every word or every sentence, or that his own version is without mistakes. Such a version could, in his opinion, only be created by Christ or the Holy Ghost, but not by a human being.[5] Thus, in spite of the superiority of his German version over all other translations, Luther does not claim that his German Bible can provide a basis for the interpretation of God's word. In this respect the humanistic

---

[1] T.R. vol. IV, no. 5009, of 21 May to 11 June 1540.
[2] L.E. vol. II, no. 449 (48–56), of 13 January 1522.
[3] E.A. vol. LXIII, p. 24.
[4] T.R. vol. V, no. 5324, of 19 October to 5 November 1540.
[5] *Summarien*..., 1531–3 (*W.A.* vol. XXXVIII, p. 16 (24–6); *E.A.* vol. LXIII, pp. 23–4).

THE INSPIRATIONAL VIEW: LUTHER

demand is fully accepted by Luther: only the original text can lead to an understanding of the truth. Therefore the study of Holy Writ in the original languages must needs be continued by those who have 'the knowledge of Christ and of God's grace'.

Luther's interpretation of the sacred text is derived from his inspiration which gave him faith and subsequently a clear perception of the meaning of Scripture without having to consult the Fathers of the Church and other expositors. This clear insight enabled him to make a new translation. For this reason I do not hesitate to call his principle of translating inspirational.

However, Luther's interpretation of the 'inspirational principle' is not identical with that postulated by Philo or Augustine. It will be remembered that Augustine contended that the ordinary philological translator is in 'servitude' to the words. The translator 'whose mind is filled with, and guided by, the divine power' is allowed not only to change the words of the original without a change of its meaning but he may also add to, or omit from, the text. This can lead to the logical conclusion that the translator may leave the original text altogether and follow his inspiration instead. But in this case he ceases to be a translator. Thus in practice a realization of this method can never be accomplished in its entirety. However much the translator may follow (or may think he follows) his inspiration, he is still bound to translate and that means that he must still render a given text. As shown above, Luther connected the actual text of Scripture with the inspiration essential for perceiving its real meaning. The translator, in Luther's view, is not a mere instrument who writes his words 'as though dictated by an invisible prompter' as Philo had maintained. The translator must have 'the knowledge of Christ and of God's grace', but besides he must possess the linguistic knowledge necessary for the understanding and interpretation of the word. This combination of inspiration and of human knowledge enables him to create a version which, though in his view not without mistakes, is yet the best translation a man can make. The translator who follows the exact wording of the original languages closely is, therefore, in 'servitude' to the words. But at the same time he will attempt to render not only the 'letter that killeth' but also the spirit. In this way he is not bound to render every word and every idiom literally and therefore is not in 'servitude' to the words. The polarity between study and inspiration must of necessity be reflected in Luther's method of translation. His method, in which he differed from earlier translators of the Bible, is outlined in the following pages.

Luther's method of translation is in full harmony with his thought and exemplifies St Augustine's saying that the translator should not be in 'servitude' to the words.

His early statements on translation, found in letters to friends, are regrettably short and even vague. They show that Luther was in agreement with the humanists that the style of a literary work, its rhetorical form, must be preserved in the rendering. On this account he criticized Scheurl, whose translation was more ornate than the original.[1] But style could not, in Luther's view, be imitated by a literal adherence to the text. The translator should not be like a 'prisoner' but should dare to change complete sentences. In 1520 Spalatin was reproached by Luther for his close imitation of the original wording in his rendering: 'figures of speech and the liveliness of sentences and of arguments can be rendered in a free translation only.' Significantly Luther adds: 'Not to speak of the difficulties involved in reproducing the spirit of the author.'[2]

These two complementary passages contain the basis of Luther's method of translation: content and form must be preserved in a free rendering, but the translator must give special consideration to the 'spirit of the author'. Luther was aware that a word-for-word translation could not reproduce the literary form and atmosphere of the original. Hebrew, Greek, Latin and German, all had peculiarities which obliged the translator to refashion many of the

---

[1] L.E. vol. I, no. 39 (1–5), of 6 May 1517 or soon after.
[2] L.E. vol. II, no. 355 (6–13), of 29 November 1520.

idioms and even rhetorical devices.¹ Latin often hindered the translator from writing a good German style.²

A real translation is the application of sayings in a foreign language to one's own language

Luther is reported to have said.³

These are the broad principles of Luther's theory of translation. He applied them to his work on the Bible with certain modifications. It will be recalled that he objected to Spalatin's renderings because they did not retain 'the spirit of the author'. This has its counterpart in his criticism of Erasmus and St Jerome for having reproduced only the dead letter of Holy Writ and in his rejection of the 'carnal' understanding of the Bible in earlier expositors. In his exegesis he believed he had found a way to the 'spiritual' interpretation of Holy Scripture. This, his theological view, is the basis of his translation. If this is kept in mind apparently contradictory statements on translation can be seen as coherent and supporting one another.

Thus there appears to be a contradiction between Luther's demand for free translation and an occasional strictly literal rendering. In these cases the impossibility of expressing the full contents of the original in the idiomatic phrases of the foreign language led Luther to prefer a word-for-word translation, even though it strained the potentialities of German. Luther himself gave on two occasions the reasons for a word-for-word translation. In Psalm lxvii. 19 (lxviii. 18) he renders the Hebrew text with 'Du hast das gefengnis gefangen' instead of the German phrase 'Du hast die gefangnen erlöset' which reads in the authorized version (lxviii. 18) 'thou hast led captivity captive'. In Luther's opinion this verse has three parallels in the New Testament: Gal. ii. 19:

---

[1] See, for example, T.R. vol. II, no. 2771 a, b, of 28 September to 23 November 1532; T.R. vol. I, no. 630, of autumn 1533; T.R. vol. I, no 1040, between 1530 and 1535; T.R. vol. V, no. 5328 and 5330, of 5 November 1540; T.R. vol. V, no. 5521, of winter 1542/3.

[2] *Sendbrief...* (W.A. vol. XXX, part 2, p. 637 (34–5); cf. pp. 636–40). *Summarien...*, 1531–3 (W.A. vol. XXXVIII, pp. 9–17).

[3] T.R. vol. II, no. 2771 a, of 28 September to 23 November 1532.

THE INSPIRATIONAL VIEW: LUTHER

'For I through the law am dead to (R.V. died unto) the law.' Rom. viii. 3: 'He...condemned sin in the flesh (*M.Gl.*: *or* for sin)' which is rendered by Luther: 'Christus hat die sunde durch die sunde verdampt.' And 2 Tim. i. 10: 'Christ Jesus who hath abolished death.'[1] Thus the theological content, and its relation to other verses which interpret this verse of the Psalm as referring to Christ, compel Luther to render word for word. This is important for the understanding of Luther's explanation of another word-for-word translation. After saying that he has preferred to retain the Greek idiomatic expression in German, he continues that a translator of the Bible needs true faith.[2] Such faith, it may be inferred, is the most essential requirement for a rendering in which the spirit of the original is to be retained.

From these two examples it can be seen that even where Luther translates word for word, he cannot be described as being in 'servitude' to the words. He does not even adhere to a specific method of translation since he sometimes renders word for word, although generally he follows the sense rather than the words. All this is possible and even necessary because it is a theologian who decides on the exact meaning of a verse and its association with other verses. Only after deciding this does he consider which method of translation to use. If it is possible to find expressions which freely render the meaning that has to be conveyed, a translation according to sense will be made. Thus Luther can say: if the angel had spoken to Mary in German, he would have used the appropriate form of address; this, and no other word, is the best translation whatever the phrase in the original may be.[3] If, however, the meaning of the original language cannot be rendered into a foreign idiom without a change of its theological meaning, a word-for-word translation can be made.

---

[1] *Summarien*...(*W.A.* vol. xxxviii, p. 13 (3–21)). Luther's translation of the last quotation is, 'Der tod ist durch Christum getödtet.' This, I assume, refers to 2 Tim. i. 10.

[2] *Sendbrief*...(*W.A.* vol. xxx, part 2, p. 640 (19–32)), literally: 'a righteous, upright, faithful, industrious, God-fearing, Christian, learned, experienced, practised heart.' Cf. above, p. 201.

[3] *Ibid.* p. 638 (21–6).

THE INSPIRATIONAL VIEW: LUTHER

At this point it can be understood why Luther deviated from the medieval tradition of Bible translation and used the method according to sense. It was not simply a logical application of the method of rendering profane literature, as practised at his time, to Bible translation, and it was not only the desire to be generally understood. It was rather his theological interpretation of the Bible which forced the method of translation upon him. He paid great attention to the language of the common man and he had the gift of writing in a clear, imaginative language. Yet however much he tried to find the right expression,[1] the ultimate intention was to make clear his theological interpretation of the text, an interpretation based on inspiration.[2]

As mentioned above, Luther was convinced of the superiority of his version over all other Bible translations. He criticized St Jerome by quoting Matthew xviii. 20: 'For where two or three are gathered together in my name, there am I in the midst of them.' In his *Table Talks* he is reported to have said:

St Jerome did as much as one person can do. No single person could have done so much. If he had had one or two associates in his work, the Holy Spirit would have been with him, as it is written: 'For where two or three are gathered together...' etc. A translator should not be alone, for the correct and appropriate words do not always occur to one individual person.[3]

This passage shows the importance attached by Luther to collaboration in translation. It is easier to find a happy turn of phrase when the translator need not rely on his own resources alone. From the very beginning of his rendering of the Bible Luther asked his friends for advice.[4] Moreover, he revised the New Testament with Melanchthon and after his return from the Wartburg received the advice of his friends on his version of the

---

[1] See, for example, *Sendbrief*...(*W.A.* vol. xxx, part 2, p. 636 (15 ff.)).

[2] For details on Luther's interpretation a paper of mine is being published in *The Journal of Theological Studies*, 1955.

[3] *T.R.* vol. I, 961, between 1530 and 1535.

[4] E.g. *L.E.* vol. II, nos. 470, of 30 March 1522; 488, of 10 May 1522; 490, of 15 May 1522; 503, of [beginning of June 1522]; 553, of [11 December ? 1522]; 556, of [12 December ? 1522].

THE INSPIRATIONAL VIEW: LUTHER

Old Testament and on his revisions. In 1531 and 1534, and from 1539 onwards special meetings[1] between Luther and his collaborators were arranged for these revisions. Minutes of these meetings have been preserved.[2] It is also possible to find evidence in Luther's works that his friends helped him in his translation. In *Eyn sendbrieff Von Dolmetzschen*, which is dated 12 September 1530, it is expressly stated that he rendered the Book of Job together with Melanchthon and Aurogallus.[3] This brings us back to 1524, the year of this translation, and it is safe to assume that he had the assistance of his friends for all parts of the Old Testament. This association can be easily explained, for Luther wrote in 1523 that the task of translating the Bible, and especially the Old Testament, was too difficult for him.

It has been mentioned that Luther never claimed to have received the understanding of every word of the Bible. He always had to ask for God's grace to give him perception of His Holy word. This perception could more easily be reached if the translator did not work alone.

The relationship between this inspirational method of translation and that based on purely grammatical considerations is of great importance. It has been said that Luther worked in two different spheres and had, in difficult sentences, to work from each of them towards the other and to come to an agreement between them.[4] Indeed in his *Table Talks* he said that the understanding of one word often opened up to him the meaning of a whole sentence. In the example adduced to prove this, he pointed out that the explanation of one Hebrew word led him to the true understanding of Psalm vi. 8.[5] His respect for the original text forced him to pursue grammatical studies.

---

[1] A vivid description of these meetings is given by M. J. Mathesius, *Historien/Von des Ehrwirdigen in Gott seligen theuren Manns Gottes/D. Martin Luthers/Anfang/Lere/Leben/ Standhafft bekentnusz seines Glaubens/vnd Sterben*... (Nürnberg, 1573), p. 151^{r-v}.

[2] Published in *WA.B.* vols. III and IV.

[3] *W.A.* vol. XXX, part 2, p. 636 (18–20).

[4] K. Holl, 'Luthers Bedeutung für den Fortschritt der Auslegekunst', *loc. cit.* vol. I (Tübingen, 1927), p. 569.

[5] *T.R.* vol. IV, no. 4149, of 25 November 1538. The Hebrew scholar Aurogallus explained the word כעס to Luther.

His opinion about grammar is clearly expressed in one of his *Table Talks* where he is reported as saying:

> When translating I always follow the rule not to fight against grammar; he who has properly recognized this, knows how to render the letter, though not the spirit.[1]

In this sentence Luther indicates the limitations of a purely grammatical translation, as he had done in his letter to Spalatin of 1520. The words 'though not the spirit' make it clear that the grammarian must fail in his version, and thus humanistic translation is condemned. Both Reuchlin and Erasmus fought for the right of the philologist to discuss the Bible and to discover the truth hidden in God's word. Luther, although not denying this right, is convinced that the philologist can understand the word only and not the spirit. His work may assist the theologian, but it is the theologian and not the philologist who will be able to discover the true meaning of Holy Writ. Some quotations may be adduced here to prove this point:

> Languages themselves do not make a theologian but they are of assistance, for it is necessary to know the subject matter before it can be expressed through languages.[2]

> It is not enough to know grammar but one must pay attention to the sense: for the knowledge of the subject matter brings out the meaning of the words.[3]

> Grammar is the knowledge of the meaning of words. Before this one must learn the meaning of the subject matter...Grammar only teaches the words which are signs for real things, as, for example, 'The righteous shall live by faith' [Rom. i. 17]. Grammar explains the meaning of 'faith', 'righteous', 'live'—but it is the sign of the highest art to defend these words against cavillers; this does not pertain to grammar but to theology.[4]

---

[1] *T.R.* vol. II, no. 2382, of 1 to 9 January 1532.
[2] *T.R.* vol. II, no. 2758 a, b, of 28 September to 23 November 1532.
[3] *T.R.* vol. IV, no. 5002, of 21 May and 11 June 1540.
[4] *T.R.* vol. II, no. 2533, a, b, of March 1522. Cf. *T.R.* vol. VI, no. 7070 of uncertain date = Scheel, pp. 176 (38)–177 (4). Cf. the agreement with *An die Ratherren*...(*W.A.* vol. XV, p. 40 (14–26)), discussed above, p. 196.

THE INSPIRATIONAL VIEW: LUTHER

God be thanked, when I understood the subject matter and knew that 'God's righteousness' meant 'righteousness through which He justifies us through righteousness freely given in Jesus Christ', then I understood the grammar. Only then did I find the Psalter to my taste.[1]

Thus the obvious conclusion is:

Grammar is necessary for declension, conjugation and construction of sentences, but in speech the meaning and subject matter must be considered, not the grammar, for the grammar shall not rule over the meaning.[2]

'The grammar shall not rule over the meaning.' This rule makes it possible to reject the grammarian if he has, in Luther's view, misinterpreted the text. The grammarian can make a grammatical rule only if his understanding of the text is correct.

It is known how Luther interpreted difficult sentences. First of all he considered whether they referred to 'grace' or 'law', to God's anger' or to 'remission of sins'. He further related the difficult passage to the context and to Christ. In this way he often came to an understanding of the meaning. These, his first considerations, were followed by an inquiry into the grammar of the sentence. He then found with the help of his friends that his interpretation could be justified by grammar or that in the case of a Hebrew text a change in the vowels only was needed, not a change in the consonants of the words.[3] There is no evidence that Luther abandoned his suggested interpretation because the grammarian did not support his view. This, however, may be accidental.

Luther did not believe that any translation of the Bible could be quoted for the purpose of theological discussion. Like the humanists, he believed that only the original could be used for the interpretation of Holy Writ, for no human endeavour could produce a translation which would replace God's word. Yet he wanted the Bible to be read by the public in the vernacular German. The resolution of the apparent logical contradiction between these two

---
[1] T.R. vol. v, no. 5247, of 2 to 17 September 1540.
[2] T.R. vol. III, no. 3794, of 25 and 27 March 1532.
[3] T.R. vol. I, no. 312 of summer and autumn 1532; vol. I, 1183, between 1530 and 1535; vol. IV, no. 5002, of 21 May to 11 June 1540; vol. v, no. 5533, of winter 1542/3.

views is of some delicacy, since of necessity it creates Bibles of different values: the original for the learned theologian who is able to interpret the text, and the translation for the congregation, who cannot arrive at an exegesis without the help of the theologian. The fact that a translation is authorized may imply that it may be used in church, or read at home. But it does not mean that it can be treated as if it were the original.

Here again, the humanists and Luther were promoting the same tendencies. New versions were devoid of theological significance and the translator could work without fear of the stigma of heresy. The thought could gain currency that, despite its sacredness, the Bible could be rendered in the same way as any literary work.

SELECT BIBLIOGRAPHY

All the bibliographical details of the primary sources are mentioned in the footnotes of this book. For this reason only secondary literature will be listed here. Since the reader will find bibliography in most of the literature enumerated below, I am referring to very few books only.

BIBLIOGRAPHIES

DOLS, J. M. E. *Bibliographie der Moderne Devotie*. 3 vols. Nijmegen, 1936–7.
SCHOTTENLOHER, K. *Bibliographie zur deutschen Geschichte im Zeitalter der Glaubensspaltung 1517–1585*. 6 vols. Leipzig, 1933–40.

CHAPTERS 1–3

BERGER, S. *Histoire de la Vulgate pendant les premiers siècles du moyen âge*. Paris, 1893.
BLONDHEIM, D. S. *Les Parlers judéo-romans et la Vetus Latina*. Paris, 1925.
COURCELLE, P. *Les lettres grecques en Occident de Macrobe à Cassiodore*. (Bibliothèque des Écoles françaises d'Athènes et de Rome, vol. CLIX.) 2e éd. Paris, 1948.
GORCE, D. *La Lectio Divina. I. Saint Jérôme et la lecture sacrée dans le milieu ascétique romain*. Paris, 1925.
HUET, PIERRE-DANIEL. *De interpretatione libri duo*. Paris, 1661.
Preface to the Reader, in the Authorized Version, 1611.
RITTER, G. *Studien zur Spätscholastik*. (S.H.A., 1921 (4. Abhandlung); 1922 (7. Abhandlung).)
SMALLEY, B. *The Study of the Bible in the Middle Ages*. Oxford, 1952.
SPICQ, P. C. *Esquisse d'une histoire de l'exégèse latine au moyen âge*. (Bibliothèque Thomiste, 26.) Paris, 1944.

CHAPTER 4

BAUCH, G. 'Die Einführung des Hebräischen in Wittenberg' (*Monatsschrift für Geschichte und Wissenschaft des Judentums*, 48 (N.F. 12) (1904), pp. 22–32, 77–86, 145–60, 214–23, 283–99, 328–40, 461–90).
GEIGER, L. *Johann Reuchlin*. Leipzig, 1871.
KLUGE, O. 'Die hebräische Sprachwissenschaft in Deutschland im Zeitalter des Humanismus' (*Zeitschrift für die Geschichte der Juden in Deutschland*, 3 (1931), pp. 81–97, 180–93).

SELECT BIBLIOGRAPHY

CHAPTER 5

The most important contribution to the knowledge of Erasmus and his contemporaries is contained in P. S. Allen's edition of Erasmus's letters. The information given in the introductions to the individual letters and in the Appendices is an invaluable aid to the student of fifteenth- and sixteenth-century humanism.

ALLEN, P. S. *The Age of Erasmus*. Oxford, 1914.
ALLEN, P. S. *Erasmus. Lectures and Wayfaring Sketches*. Oxford, 1934.
P. S. & H. M. ALLEN and H. W. GARROD, edd. *Opus Epistolarum Des. Erasmi Roterodami*. 11 vols. Oxford, 1906–47.
Bibliotheca Erasmiana. *Bibliographie des œuvres d'Érasme*. 6 vols. Ghent, 1897–1906.
Bibliotheca Erasmiana. *Listes sommaires*. Ghent, 1893.
MANGAN, J. J. *Life, Character and Influence of Desiderius Erasmus of Rotterdam*. 2 vols. London, 1927.
MURRAY, R. H. *Erasmus and Luther: Their Attitude to Toleration*. London, 1920.
PFEIFFER, R. *Humanitas Erasmiana. Studien der Bibliothek Warburg*, vol. XXII. Leipzig, Berlin, 1931.
PINEAU, J. B. *Erasme: sa pensée religieuse*. Paris, 1924.
RENAUDET, A. *Érasme. Sa pensée religieuse et son action d'après sa correspondance (1518–1521)*. Paris, 1926.

CHAPTER 6

BÖHMER, H. *Luther and the Reformation in the Light of Modern Research*. Translated by E. S. G. Potter. London, 1930.
HOLL, C. *Gesammelte Aufsätze zur Kirchengeschichte*, vols. 1 and 3. Tübingen, 1927, 1928.
KAWERAU, G. 'Luthers Schriften nach der Reihenfolge der Jahre verzeichnet, mit Nachweis ihres Fundortes in den jetzt gebräuchlichen Ausgaben.' 2. Aufl. durchgesehen by O. Clemen. *Schriften des Vereins für Reformationsgeschichte*, 47, Heft 2 (Nr. 147), pp. 163–206. Leipzig, 1929.
RUPP, G. *The Righteousness of God. Luther Studies*. The Birkbeck Lectures in Ecclesiastical History, 1947. London, 1953.
SCHEEL, O. *Martin Luther. Vom Katholizismus zur Reformation*. 3rd ed. Tübingen, 1921, 1930.
VOGELSANG, E. *Die Anfänge von Luthers Christologie nach der ersten Psalmenvorlesung*. Arbeiten zur Kirchengeschichte, vol. 15. Berlin and Leipzig, 1929.
WOLFF, O. *Haupttypen der neueren Lutherdeutung*. Tübinger Studien zur systematischen Theologie, Heft 7. Stuttgart, 1938.

BIBLE INDEX

OLD TESTAMENT

Book of Judges, 130
Books of Kings, 130
Psalms, 178–9; Ps. ii. 10, pp. 178, 179, 200 n.; iv. 4, p. 178; vi. 8, p. 209; l. 5 (li. 5), p. 126, 126 n.; l. 18 (li. 18), p. 74; lix (lx), p. 175; lxvii. 19 (lxviii. 18), p. 206; lxxii. 16–17 (lxxiii. 16–17), p. 176; xci. 15 (xcii. 14), p. 126, 126 n.; ci. 24 (cii. 24), p. 73; cix. 2 (cx. 2), pp. 85–6; cx. 10 (cxi. 10), p. 180 n.; cxviii. 10 (cxix. 10), p. 180; cxxix. 4 (cxxx. 4), p. 73; cxxxviii. 14 (cxxxix. 15), p. 178 n.; cxl. 6 (cxli. 6), pp. 77, 80
Proverbs, 84 n., 123, 123 n.
Isaiah, 42
Jeremiah, 42
Ezekiel, 82
Amos, vii. 14, pp. 176, 177
Jonah, 38
Habakkuk, ii. 4, p. 169

NEW TESTAMENT

Gospel of St Matthew, 27, 139 n.; Matt. i. 19, p. 144 n.; iv. 10, pp. 133 n., 151; vi. 9–13, p. 145; vi. 13, pp. 143–4; vii. 24, p. 133 n.; xi. 30, pp. 145 n., 147; xvii. 5, p. 150 n.; xviii. 20, pp. 202, 208; xxviii. 8, p. 133 n.; xviii. 24, 133 n.
Gospel of St Mark, 139 n.; Mark, ix. 34, p. 176
Gospel of St Luke, 139 n.; Luke, i. 26–7, pp. 48–50; i. 28, pp. 51, 53, 54–5; i. 31, p. 54; xxiv. 45, p. 175
Gospel of St John, 14, 139 n.; John, i. 1, p. 142 n.; i. 14, p. 54
Acts, 139 n.; Acts vi. 8, p. 53
Epistles of St Paul, 117, 122, 128, 131, 134, 152, 153, 159, 174, 177, 179, 185, 185 n., 189 n., 191
Epistle to the Romans, 189 n.; Rom. i. 4, p. 151; i. 17, pp. 168, 168 n., 169, 210); ii. 4, p. 190 n.; vii. 14, p. 174; viii. 3, p. 207; viii. 33, p. 186 n.; ix, p. 186; ix. 15, p. 186 n.; xii. 2, p. 191; xiv. 9, p. 188 n.; xiv. 14, p. 188, 188 n.; xiv. 16, p. 188 n.; xiv. 19, p. 189 n.; xiv. 22, p. 188 n.; xv. 26, p. 188 n.
Epistles to the Corinthians: 1 Cor. ii. 8, p. 179; ii. 9, p. 179; ii. 16, p. 179; iv. 3, p. 153 n.; iv. 3–4, p. 133 n.; vi. 20, p. 133 n.; ix. 13, p. 134 n.; xii. 4–11, p. 53; xv. 31–2, p. 133 n.
Epistle to the Galatians, 187, 189 n.; Gal. i. 6, p. 188 n.; i. 10, p. 170; ii. 2, p. 186 n.; ii. 10, p. 188 n.; ii. 16, p. 188 n.; ii. 19, pp. 206–7; ii. 20, p. 188 n.; v. 4, p. 187 n.; v. 9, p. 187 n.; v. 10, p. 190 n.
Epistle to the Ephesians, 153, 153 n.
Epistles to the Thessalonians: 1 Thess. i. 8, p. 134 n.
Epistles to Timothy: 2 Tim. i. 10, p. 207, 207 n.
Epistle to the Hebrews, 189 n.; Heb. ii. 10, p. 189 n.; v. 11, p. 188 n.; vi. 7, p. 189 n.; viii. 13, p. 188 n.; ix. 1, p. 188 n.; xiii. 24, p. 153, 153 n.
St Peter, 140
St John, 140; 1 John ii. 15–16, pp. 140, 153 n.
Revelation, 35, 139 n.

EDITIONS AND TRANSLATIONS, COMPLETE BIBLES

American Revised Standard Version, 2, 6
American Standard Revision, 6
Authorized Version, 1, 2, 3, 4, 5, 6, 48, 49, 51, 53 n., 73, 74, 78 n., 151, 188 n., 207
Bishops' Bible, 5
Complutensian, 47, 92 n., 94 n., 164
Cranmer, 188 n.
Douay edition, 9 n., 176, 176 n.
Geneva edition, 55, 188 n.
Knox, Mgr, 3; *see also* General Index, *s.v.*
Revised Version, 2, 5, 6, 48, 49, 179 n., 188 n., 207
Rheims edition, 9 n., 55, 77, 176, 176 n., 188 n.
Tyndale, 14, 15, 128, 158, 188, 188 n.
Vetus Latina, 26, 31, 38, 39, 40, 156

BIBLE INDEX

Vulgate, x, 46, 53 n., 154
  is declared the authentic text, 10, 10 n., 44, 44 n., 159–60, 159 n., 160 n.; importance of the V. as authoritative version, 45; cannot contain errors (Dorp), 163–4; replaces all other texts of Bible, 45; is the basis of biblical translation, 11; interrelation between the text of the V. and interpretation, 48–50, 51–2; Colet uses the V., not the original text, 120; is the best translation in existence (Luther), 202, 203; Luther relies on the V., 176, 177
  the humanists do not wish to replace the V., 62, 157–8, 160–1; style of V. criticized, 4, 85, 133, 164; its position endangered by Erasmus's New Testament, 142, 143, 157–8, 159–60; criticized by Erasmus, 126, 150, 160–2; criticized by Reuchlin, 77, 78, 78 n., 85, 137; criticized by Valla, 133–4; Luther's reliance on the V. is waning, 186, 187, 190; cannot replace the original text (Reuchlin), 74, 84–5, 86; Crastonus compares the texts of the Psalms, 106–7; Erasmus compares the Greek text with the V., 126, 126 n., 137–8, 143–4; Reuchlin compares the Hebrew text with the V., 74, 86; differences between Greek text and V., 121, 126, 126 n., 133; differences between the Hebrew text and the V., 69; Reuchlin prefers the Hebrew and Greek texts to the V., 71, 72; the Latin text causes wrong interpretation (Erasmus), 150–2; mistakes in the V. (Reuchlin), 78, 78 n., 79; mistakes in the V. (Erasmus), 136, 160; cannot be the basis of biblical translation, ix; cannot be the basis of biblical interpretation (Erasmus), 160–1; its text is corrupt, 11–12, 73, 105, 136, 143; emendation of the V. with the help of Greek manuscripts (Erasmus), 145; attempt to restore the text of the V. (Erasmus), 145, 146, 154, 157, 201
  printed editions of 1498 and 1502, 47–50; Complutensian, 47
Wyclif, 51, 55, 188 n.

EDITIONS AND TRANSLATIONS, OLD TESTAMENT

Aquila, 28 n., 42 n.
Hexapla, 26
Psalms, Crastonus, Johannes, 106, 107, 126 n.; Giustiniano, Agostino, 77 n., 94 n.; Kittel, R.–Kahle, P. E., 73 n.
Septuagint, 1, 15, 17, 22, 26, 27, 28, 28 n., 29, 30, 31, 32, 33, 37, 38, 39, 40, 41, 42, 73, 74, 156, 201, 202
Symmachus, 28 n., 73 n.
Theodotion, 28 n., 73 n.

EDITIONS AND TRANSLATIONS, NEW TESTAMENT

Erasmus, 142–57
Harwood, Edw., 1, 4–5
St Jerome's revision of the New Testament, 26
Moffatt, J., 2
Worsley, John, 3

GENERAL INDEX

Abraham aben Esra, 77
Acts, *see* Bible, New Testament, Acts
Adrianus, Matthaeus, 66 n.
Agricola, Rudolph, 57, 71 n.
Albertus the Great, 124
Aldus Manutius, *see* Manutius, Aldus
Alexandria, 18, 33; *see also* Philo of Alexandria
Allen, P. S., 56 n., 93 n., 94 n., 95 n., 98 n., 100 n., 107 n., 114 n., 121 n., 123 n., 124 n., 125 n., 128 n., 132 n., 139 n., 140 n., 142 n., 144 n., 162 n., 163 n.
Ambrose, 49, 128, 134, 156, 157, 188 n., 200
American Revised Standard Version, *see* Bible, Editions and Translations, Complete Bibles
American Standard Revision, *see* Bible, Editions and Translations, Complete Bibles
Amos, *see* Bible, Old Testament, Amos
Andrew of St Victor, 69 n.
Anselm of Canterbury, 146
Aquila, *see* Bible, Editions and Translations, Old Testament
Archbishop of Westminster, *see* Westminster, Archbishop of
Aristeas Letter, 17-21, 22, 24, 26, 33
Aristotle, 75, 78, 181
Arnold of Tungern, 89
Arundel, Archbishop, 13 n.
Augustine, St, of Hippo, xi, 29 n., 126 n., 128, 134, 135, 156, 157, 194, 197
 on translation, 42-3, 72, 72 n., 205; on the task of the translator, 42-3; on an inspired translation, 39, 40-3; follows Philo, 40; influenced by Origen?, 42 n.; mistrusts human understanding, 43; view on the Septuagint, 37, 38, 39, 40, 41, 42; on differences between the Hebrew Bible and the Septuagint, 41-2; on the Hebrew text of the Old Testament, 37, 38, 40; admits no mistake in Vulgate, 164; reports on disturbances because of a new translation of the Bible, 4, 38; favourable to Jerome's revision of the New Testament, 38-9; opposed to Jerome's translation of the Old Testament, 37, 38, 40-1, 42, 44
Augustine, views on
 quoted by Erasmus on the use of pagan literature, 98, 102; on inspiration, 99; on the Greek text of the New Testament, 105, 137, 137 n., 143; on grammar, 106; in support of his translation of the New Testament, 146, 151, 151 n.; criticized by Erasmus, 150; for solecisms, 155; Erasmus recognizes corruptions in the manuscripts of Augustine, 145 n.
 and Luther, similarity of their views and Luther's dependence on Augustine, 177, 178, 179, 189, 196, 200, 200 n.; Luther's views are confirmed in Augustine's works, 168, 168 n., 178, 192; Luther disagrees with Augustine, 201; Luther disagrees with Augustine on the meaning of the word 'inspired' translator, 204; criticized by Luther, 185; superior to Jerome according to Luther, 198, 199-200; criticized by Reuchlin, 77-8, 78 n.
Aurifaber, Johann A., 202 n.
Aurogallus, 209, 209 n.
Authorized Version, *see* Bible, Editions and Translations, Complete Bibles

Barclay, Alexander, 58 n.
Bardenhewer, O., 142 n.
Bardy, G., 36 n.
Basil, 134, 157
Bataillon, M., 163 n.
Bauch, G., 66 n.
Bebel, Henricus, 58
Bede, 146
Behaim, John, 66, 67, 67 n.
Belloc, H., 3
Beumer, J., 50
Biel, Gabriel, 52-5, 58
Bishops' Bible, *see* Bible, Editions and Translations, Complete Bibles
Blatt, F., 36 n.
Blondheim, S., 29 n.
Bludau, A., 142 n., 153 n., 163 n.
Boecking, E., 86 n.

GENERAL INDEX

Böhmer, H., 172 n., 176 n., 177 n., 180 n.
Boeschenstein, Johannes B., 68 n.
Bomberg, 80, 90
Bowie, W. R., 2 n.
Brant, Sebastian, 58, 58 n., 59, 76, 180
Brethren of the Common Life, 57, 103
Bretschneider, C. G., 68 n.
Budaeus, 93
Buhl, F., 73 n.
Burgos, Paul of, *see* Paul of Burgos

Campbell, W. E., 14 n., 60 n.
Cavallera, F., 32 n., 35 n., 48 n.
Charles, R. H., 17 n.
Christ, K., 68 n., 80 n.
Chrysostom, 92 n., 134, 157
Church of Scotland, 3 n.
Cicero, 32, 92 n., 96, 109, 125
Claudian, 96
Clement V, *see* Popes, Clement V
Cohn, L., 22 n.
Colet, Sir Henry, 139
Colet, John, his life and education, 109; his character and his austerity, 110; lectures at Oxford, 109–10
  as an interpreter of the Bible, 109–10, 120, 121; understanding of the Bible 'by grace alone', 113–14; attitude towards the Vulgate, 120; attitude towards a new translation of the Bible, 120, 139, 139 n., 144, 162, 162 n.
  and Erasmus, 108–21; exchange of letters, 95, 116–18, 122, 131, 132, 132 n.; influence on Erasmus, 111, 114, 120, 132; agreement between them, 111, 114–16, 128; disagreement between them, 112–14; tension between them, 117; invitation to Erasmus to lecture at Oxford, 114, 117; Erasmus's promise to come to Colet's aid, 117, 118; his view on Erasmus, 108–9, 162, 162 n.; Erasmus's view on Colet, 110–11, 112, 116
  opposed to heathen authors, 113, 129; attitude towards languages, esp. Greek, 120; mistrusts Reuchlin's cabbalistic studies, 87, 87 n., 114; studies the Schoolmen, 110; dislikes the Schoolmen, 111, 112, 152 n.; hates solecisms, 111
Collin, 88, 89
Colson, F. H., 22 n.

Complutensian Bible, *see* Bible, Editions and Translations, Complete Bibles
Corinthians, *see* New Testament Bible, Epistles to the Corinthians
Council of Trent, 10, 10 n., 11, 60, 159, 160, 160 n.
Council of the Vatican, 10 n.
Council of Vienne, 127, 134
Courcelle, P., 32 n., 34 n., 36 n., 37 n., 43 n.
Cranmer, *see* Bible, Editions and Translations, Complete Bibles
Crastonus, Johannes, *see* Bible, Editions and Translations, Old Testament, Psalms
Cuendet, G., 36 n.
Cyprian, 113, 146, 156, 157, 185
Cyril, 157

Daiches, D., 65 n.
Damasus, *see* Popes, Damasus
Dante, 159 n.
Deanesly, M., 13 n.
Delitzsch, F., 163 n.
Demetrius Phalereus, 18
Denifle, H., 168 n.
Dietrich, Veit, 202 n.
Dionysius the Areopagite, 113
Dorp, M., 163, 164, 164 n.
Douay Bible, *see* Bible, Editions and Translations, Complete Bibles
Drummond, R. B., 144 n.
Duns Scotus, 55, 102, 110, 124, 129, 151

Ebrardus, 97 n.
Eck, Johannes, 90 n.
Edie, W., 145 n., 163 n.
Edinburgh Review, 5, 5 n.
Eleazar, high priest, 18
Ennius, 96
Ephesians, Epistle to the, *see* Bible, New Testament, Epistle to the Ephesians
Erasmus, ix, xi, 72 n., 80 n., 90, 90 n., 91, 172; are his letters sincere?, 101, 103–4, 118
  limitations of our knowledge of Erasmus's thought, 95
Erasmus's attitude towards theology
  respects the authority of the Church, 146, 161, 166; on the philosophy of Christ and its depravation, 147–50, 158, 161–2; dislikes the religiosity of the Brethren of the Common Life and the mysticism of the *Imitatio Christi*, 103
  distinguishes between 'old' and 'new'

218

Erasmus's attitude towards theology (*cont.*)
theology, 102–4; criticizes 'new' theology, 102, 115–16, 129, 149; praises 'old' theology, 116, 125–6, 129, 149; return to the early Fathers of the Church, 103, 115, 116, 120, 125–6, 129, 149; critical of the Fathers of the Church, 149–50; uses Augustine for the defence of humanism, 98; quotes Augustine's view on inspiration, 99; critical of Augustine, 155; and Jerome, Er. uses Jerome for the defence of his views, 98, 136, 137, 153; prepares commentaries on Jerome's work, 123–5; praises Jerome who combined sacred and profane *literae*, 124–5; praises Origen, 128–30, 130 n., 131; follows scholastic tradition, 57; opposed to Schoolmen, 14, 100–1, 102, 115, 116, 124–5, 129, 141, 148, 151; on Thomas Aquinas, 134, 151–2; on 'barbarians', 97–8, 97 n., 98–9, 118; quotes Decreta of the Church for the defence of his view, 127, 134, 137, 137 n.; parts of his work condemned, 159–60

Erasmus's humanism
his vocation, 94, 99; his reading lists of 1489, 95–6; on the importance of a good style, 100, 125, 155; is a grammarian not a theologian, 78, 140–1; demands freedom to criticize, 133–4, 150; the writings of pagan authors should be studied, 98, 101–2, 104, 105, 122, 131–2, 138–9; the knowledge of proverbs is important for biblical studies, 121–3, 132; does not know the significance of Greek for theology, 105–6, 120–1
his teachers in Greek, 93, 93 n.; learns Greek, 99–100, 121–2, 123–4; on the importance of Greek, 99–100, 105, 120, 124, 126–8, 131–2, 137–8, 145, 150–1, 152
combines humanism (*literae*) and theology, 94, 104, 122–4, 128–30, 131–2, 138, 140–2, 153–4; difficulty of combining humanism and theology, 115, 118–19, 120, 130; humanists before Er. who combine humanism and theology, 104–5, 122–3; his ignorance of the implications of his thought on theology, 103–4, 130; relations between philology and theology, 135–6, 140–1, 152–4; demands a new type of education, 141, 152; on creating a new civilization, 99

Erasmus and the Bible
Bible studies occupy a central place in his life, 63, 161–2; mistrust in the Greek text of the New Testament, 105–6; his edition of the New Testament contains the first Greek text of the New Testament, 94; compares the Greek text of some Psalms with the Latin Vulgate, 126, 126 n.; the philologist should translate the Bible, 134–5; attacks the inspirational principle of translation, 135–6, 155–6; the philologist should reconstitute the text of the Bible, 136; translation of the New Testament of 1505, 139–40, 139 n., 140 n.; New Testament of 1516, 142–57, 142 n.; his method of interpretation, 151, 152–4; his method of textual criticism, 143–6; technique of his biblical translation, 154–7; applies the principles of literary translation to the Bible, 1; distinguishes between literal and allegorical exegesis, 127–30, 131–2; defends his changes of the Vulgate text, 146; justifies his biblical translation by reference to Fathers, 146; his 'New Testament' is not competing with Vulgate, 157–8, 160–1; criticizes biblical translations, 127–8; criticizes the Vulgate, 4, 126, 150, 154, 160–2; his attitude towards commentators of the Bible, 147–52; the Bible for the people, 158–9; the Bible is the only authority, 146; the Bible is the centre of the new civilization, 141

Erasmus, friends and foes
and Colet, 108–21; Er.'s biographical sketch of Colet, 110–12; personal friendship between them, 108–12; Colet invites Er. to lecture at Oxford, 114; Er. refuses Colet's invitation, 116–17, 118–19; Colet's influence on Er., 110, 120–1; similarity of outlook between Er. and Colet, 111–12, 115–16, 128, 139; differences between Er. and Colet, 112–14; his promise to Colet and its fulfilment, 117–18, 125–6; Colet approves a new biblical translation by Er., 139; his New Testament praised by Colet, 162; does not know Cras-

GENERAL INDEX

Erasmus, friends and foes (*cont.*)
    tonus's edition of the Psalter, 106–7; and Gaguin, 101–2, 105, 108, 109 n.; his work used by Luther, 167, 186, 187, 188, 188 n., 189, 189 n.; difference between Er. and Luther, 184, 190–2, 196–201, 202–3, 206, 210; his wit criticized by Luther, 183; defended by Thomas More, 59, 139, 162; attitude towards Reuchlin, 87, 87 n.; similar tendencies between Er. and Reuchlin, x, 78, 78 n., 86, 86 n., 88 n., 94, 126, 127, 127 n., 137–8, 150, 154, 210; critical of Reuchlin's cabbalistic studies, 87; defends Reuchlin against Hochstraten, 87; praises Laurentius Valla, 96–7; influenced by Laurentius Valla's theory of language, 96, 116, 152; publishes Valla's *Adnotationes*, 132; importance of Valla's *Adnotationes* for Er., 106, 134, 158
    opposition against his New Testament, 162–6; by Dorp, 163–4; Latomus, 165; Zúñiga (Stunica), 164
    see also Bible, Editions and Translations, New Testament

Euripides, 141
Eusebius, 26, 28, 29, 35 n.
Ezekiel, *see* Bible, Old Testament, Ezekiel

Faber Stapulensis, *see* Lefèvre, Jacques
Farrar, F. W., 47 n.
Faulkner, J. A., 162 n.
Ficker, J., 175 n., 178 n., 186 n., 188 n., 189 n., 197 n.
Fisher, Christopher, 132
Fisher, John, 80 n.
Fisher, Robert, 111
Foxe, J., 13 n.
Freier, M., 189 n.
Freimann, A., 66 n., 67 n.
Fretela, 34, 34 n.
Friedlaender, G., 90 n.
Froben, Hieronymus, 142 n.

Gaguin, Robert, 101, 102, 104, 105, 108, 109 n., 119
Galatians, *see* Bible, New Testament, Epistle to the Galatians
Geiger, A., 65 n., 67 n., 70 n., 71 n., 73 n., 76 n., 78 n., 80 n., 86 n., 87 n., 90 n.
Gelasius, *see* Popes, Gelasius
Geldenhouwer, Gerard, 56 n.

Gellius, 100 n.
Geneva edition of the Bible, *see* Bible, Editions and Translations, Complete Bibles
Geoffrey de Vino Salvo (Vinsauf), 96
Giustiniano, Agostino, *see* Bible, Editions and Translations, Old Testament, Psalms
Gospel of St John, *see* Bible, New Testament, Gospel of St John
Graf, K. H., 173 n., 178 n.
Gratian, 58
Gregory the Great, *see* Popes, Gregory the Great
Gregory of Nazianzus, 134
Grey, Thomas, 102, 103
Gudenus, V. F., 13 n.

Habakkuk, *see* Bible, Old Testament, Habakkuk
Haller, J., 52 n.
Hamel, A., 172 n.
Hardt, H. von der, 72 n., 74 n., 75 n., 77 n., 79 n., 80 n., 85 n.
Harnack, A., 169 n., 183 n.
Hartfelder, K., 167 n., 194 n.
Harwood, Edward, *see* Bible, Editions and Translations, New Testament
Hauck, A., 13 n., 36 n., 69 n., 70 n.
Hebrews, Epistle to the, *see* Bible, New Testament, Epistle to the Hebrews
Hefele, C. H., 10 n., 159 n.
Hegius, 93
Hendriks, O., 163 n.
Henry of Bergen, 101
Hermonymus of Sparta, 93, 93 n.
Hexapla, *see* Bible, Editions and Translations, Old Testament
Hilary, St, 79, 146, 157
Hirsch, E., 172 n., 186 n.
Hirsch, S. A., 70 n.
Hochstraten, Jacob, 87, 89
Höpfl, H., 10 n., 145 n., 159 n.
Holl, K., 168 n., 172 n., 180 n., 200 n., 209 n.
Homer, 34, 177
Horace, 70, 96
Horawitz, A. D., 68 n.
Hugo, Cardinal, 146
Huguitio, 97 n.
Hummelberg, Michael, 93
Hyma, A., 96 n., 97 n., 98 n., 99 n., 100 n., 101 n., 102 n., 104 n., 113 n., 124 n., 128 n., 131 n.

GENERAL INDEX

Isaac, F., 48 n.
Isaacs, J., 5 n.
Isaiah, *see* Bible, Old Testament, Isaiah

Jacob ben Jehiel Loans, 65
James I, *see* Bible, Authorized Version
Jeremiah, *see* Bible, Old Testament, Jeremiah
Jerome, St, ix, xi, 46, 51, 74 n., 89, 112, 128, 134, 137 n., 194
  doubts about the inspirational theory of translation, 27–8; inspirational translation and prophecy, 30; different conception of inspirational theory from Philo, 29–30; the inspired version does not replace the original, 30; no inspired translation exists, 32–3; difference between translation and prophecy, 25, 32, 37
  uses the philological method of Bible translation, 31–2, 37; only the original text of the Bible can be the basis of a translation, 33; doubts about the correct rendering of the Bible, 27–8; on the task of a translator, 27, 29, 34–7; different method of translation for the Bible and profane books, 34–7; conflicting statements about the method of translation, 35; importance of style in translations, 26–7, 29, 34; retains the old rendering against his will, 36; on difficulties of Bible translation, 27, 36; his view on the Septuagint, 26–8, 29, 31, 32–3; on the 'legend' of the origin of the Septuagint, 32–3; is asked to revise the *Vetus Latina*, 26
  his philological method of translation is censured by St Augustine, 37–9, 40–4; importance of his translation in the Middle Ages, 45, 50; his Vulgate becomes the authentic text, 44, 44 n.
Jerome, views on
  quoted by Colet, 120; Crastonus on Jerome's Psalm translation, 106; influences Erasmus, 105, 123; referred to by Erasmus, 98, 134, 135–6, 136 n., 143, 144, 153, 153 n., 156, 157; praised by Erasmus for his erudition and style, 124–5, 127 n.; publication of his works by Erasmus, 124–5, 125 n., 126, 127; Luther prefers Jerome to Augustine, 189; Luther prefers Augustine to Jerome, 196, 198, 199–200; censured by Luther, 185, 202–3, 208; censured together with Erasmus by Luther, 202–3, 206; his commentaries useful for learning the 'simple story', according to Luther, 199–200; the Vulgate not translated by Jerome according to Luther, 187; his translation of the Psalms printed by Lefèvre, 177; influences Reuchlin, 71, 76, 150; quoted by Reuchlin, 71–2, 83, 83 n., 88, 88 n.; criticized by Reuchlin, 77
  see also Bible, Editions and Translations, New Testament
Jethro, 135
John the Baptist, 153
John, St, *see* Bible, New Testament, Gospel of St John
John, St, *see* Bible, New Testament, Epistles, St John
John, St, Revelation of, *see* Bible, New Testament, Revelation
Jonah, *see* Bible, Old Testament, Jonah
Josephus, 33
Judaea, 22
Judges, Book of, *see* Bible, Old Testament, Book of Judges
Juvenal, 96

Kahle, P. E., 17 n.; *see also* Bible, Editions and Translations, Old Testament, Psalms, *s.v.* Kittel, R.–Kahle, P. E.
Kawerau, G., 70 n., 172 n.
Kenyon, F. G., 163 n.
Kimhi, David, 77
Kings, Books of, *see* Bible, Old Testament, Books of Kings
Kittel, R., *see* Bible, Editions and Translations, Old Testament, Psalms
Kluge, O., 65 n., 69 n., 73 n., 76 n., 93 n.
Knod, G., 71 n.
Knox, Mgr, 3, 3 n., 4 n., 8 n., 12, 12 n.; *see* Bible, Editions and Translations, Complete Bibles
Kroker, E., 202 n.
Kropatscheck, F., 13 n., 169 n.

Lactantius, 146
Lang, Johannes, 167, 198
Lascaris, C., 66, 92
Latomus, Jacob, 165, 165 n.
Lefèvre, Jacques (Faber Stapulensis), 91, 172, 172 n., 173–4, 175, 177, 178, 179, 186, 186 n., 187, 187 n., 188 n., 189, 197
Leo X, *see* Popes, Leo X

GENERAL INDEX

Leo XIII, *see* Popes, Leo XIII
Levita, Elias, 73 n.
Lewis, C. S., 5
Lilje, H., 195 n.
Lilly, W. S., 70 n., 71 n.
Livy, Titus, 129, 130
Locher, Jacob, 58 n.
Lucan, 96
Lucas of Bruges, Francis, 145 n.
Luke, St, *see* Bible, New Testament, Gospel of St Luke
Lupton, J. H., 109 n., 111 n., 112 n., 113 n.
Luther, Martin, ix, 59, 81 n.
   'inspirational' principle, 195–6, 204–5; his 'inspiration', 167–72; his judgement on his 'inspiration', 167–71; 'inspiration' as the basis of exegesis, 169–71, 176, 181–3, 185, 191, 193, 195–6, 197–8, 199–200, 202, 208; his 'inspiration' is the basis of his criticism of Erasmus's New Testament, 172; his 'inspiration' and *Lecture on the Psalms*, 172–3, 176, 178; faith is essential for the understanding of Scripture, 200–1; God's grace is necessary for the understanding of the Bible, 169–71, 175–6, 180, 184–5, 191, 195–6, 197, 199–200; only Christians can understand the Bible, 175–6, 200–1, 202; on understanding the Bible, 179–81; how to attain wisdom, 179–81; rejects or mistrusts human wisdom, 179, 180–1, 184–5, 192, 197–8, 199–200
   and the fourfold exegesis of the Bible, 177, 200; and the original text of the Bible, 176, 186, 187, 189, 190, 194, 195, 203–4, 209; uses the Greek New Testament, 186, 187; his Bible translation heretic according to Thomas More, 14–15; view about earlier translators of the Bible, 201, 202, 203; criticizes the Septuagint, 201–2; and the Vulgate, 177, 186, 187, 202–3, 208; no biblical translation is final, 211–12; L.'s Bible translation influences Tyndale's, 14, 15; possesses the Hebrew Bible printed at Brescia in 1494, 68 n.
   respects the authority of the Church, 181–2, 183; critical of the Fathers of the Church, 185, 189, 194
   and Augustine, similarity of their views and Luther's dependence on Augustine, 177, 178, 179, 189, 196, 200, 200 n.; finds his views confirmed in Augustine's works, 168, 168 n., 178, 192; critical of Augustine, 185; disagrees with Augustine, 201; disagrees with Augustine on the meaning of the term 'inspired' translator, 204; believes Augustine to be superior to Jerome, 198, 199–200; and St Jerome, 177, 185, 187, 189, 194, 196, 199, 200, 201, 202, 203, 206, 208
   view on grammar and philology, 186–92, 193–4, 197, 209–11; his knowledge of Hebrew, 178; attitude towards the Hebrew text of the Old Testament, 176, 189; makes use of Hebrew grammar, 177, 178–9, 178 n.; attitude towards Hebrew commentators, 175, 176, 177, 201–2; on the corruption of the Hebrew language, 201–2
   and humanism, 167, 183, 186–94, 196–201, 212; and Erasmus, 186–9, 190–2, 196–200, 203, 206; makes use of Erasmus's philological observations, 186–9, 196; difference between Erasmus and Luther, x, xi, 184, 190–2, 196–201, 202–3, 206, 210; writes against Erasmus, 57; and Melanchthon, 176, 194, 195; and Reuchlin, 90, 178; Reuchlin is opposed to Luther, 86; on the importance of languages, 193–5, 198–9
   on penitence, 190–1
   contrast between philosophy and theology, 179–81, 184–5; opposed to philosophy, 184; attitude towards Aristotle, 180, 184
   Lecture on the Psalms, 172–83; his sources, 172–3, 177, 178–9; and Jacques Lefèvre, 172, 174–7; principle of biblical interpretation, 169–71, 174–5, 177
   and Scholasticism, 181, 183, 184–5
   on translation, 204, 205–12; accuses the Universities, 90; claims Wessel Gansfort as a predecessor, 56
Lyra, Nicolaus of, *see* Nicolaus of Lyra

Mackinnon, J., 172 n.
Madan, F., 92 n.
Mangan, J. J., 115 n.
Mann, M., 163 n., 172 n., 173 n.
Manutius, Aldus, 66, 92
Mark, St, *see* Bible, New Testament, Gospel of St Mark

GENERAL INDEX

Marschalk, Nicolaus, 66
Martial, 96
Marx, A., 66 n., 67 n.
Mathesius, M. J., 209 n.
Matthew, St, *see* Bible, New Testament, Gospel of St Matthew
Maximilian, Emperor, 88
Meecham, H. G., 17 n.
Meghen, Peter, 139, 139 n.
Meissinger, K. A., 172 n., 175 n., 186 n., 192 n.
Melanchthon, 68 n., 167, 194, 194 n., 195, 203, 208, 209
Mestwerdt, P., 133 n.
Miller, E. W., 56 n.
Mirbt, C., 13 n.
Moffatt, J., *see* Bible, Editions and Translations, New Testament
More, Thomas, 14, 15, 58–60, 139, 139 n., 162, 162 n.
Moses ben Gabirol, 77
Mozley, J. F., 128 n.
Müntzer, Thomas, 193, 200
Mutian, Conrad, 139, 139 n., 178 n.

Nestle, E., 145 n., 163 n.
Nichols, F. M., 97 n., 98 n., 102 n., 103 n., 108 n., 109 n., 111 n., 114 n., 115 n., 116 n., 118 n., 122 n., 131 n., 132 n., 135 n., 136 n., 138 n., 141 n., 158 n.
Nicolaus of Lyra, 48, 49, 51, 64 n., 69, 73, 76, 88, 89, 134, 135, 150, 176, 189, 197

Occam, William of, 52, 58, 91
Old Latin Version, *see* Bible, Editions and Translations, Complete Bibles, Vetus Latina
Origen, 26, 42 n., 82 n., 83, 83 n., 113, 120, 128, 130, 130 n., 131, 157, 185
Ovid, 96

Pahl, Th., 178 n.
Papia, 97 n.
Paul, St, *see* Bible, New Testament, Epistles of St Paul
Paul of Burgos, 69, 73, 176
Pauli, Johannes, 65, 65 n., 66
Pellican, Conrad, 56 n., 65, 65 n., 66, 66 n., 67, 67 n., 69–70
Perles, J., 65 n.
Persius, 96
Peter, St, *see* Bible, New Testament, Epistles, St Peter

Peter d'Ailly, 56
Peter Lombard, 52, 53, 151, 153, 177
Petrarch, 97, 159 n.
Pfefferkorn, 88
Pfeiffer, R., 94 n., 96 n., 97 n., 104 n., 130 n., 133 n.
Philadelphus, *see* Ptolemy II
Philelphus, 97
Philo of Alexandria, 21–5, 26, 29, 30, 32, 32 n., 40, 81 n., 196, 204
Pico della Mirandola, 64, 80, 80 n., 91, 128 n.
Pijper, F., 165 n.
Pineau, J. B., 114 n., 146
Pius XII, *see* Popes, Pius XII
Plato, 81, 109, 111
Plautus, 117
Plomer, H. R., 92 n.
Plotinus, 109
Polman, P., 160 n.
Polydore, *see* Vergil
Popes, Clement V, 79, 87; Damasus, 26; Gelasius, 88; Gregory the Great, 51, 185; Leo X, 94, 142, 142 n.; Leo XIII, 9, 9 n., 11, 159 n.; Pius XII, 7–8, 8 n., 9–10, 13, 13 n.
Preuss, H., 172 n., 182 n., 196 n., 197 n.
Priscian, 74
Proctor, R., 47 n., 92 n.
Propertius, 96
Proverbs, *see* Bible, Old Testament, Proverbs
Prudentius, 146
Psalms, *see* Bible, Old Testament, Psalms
Ptolemy II, Philadelphus, 17, 18, 19, 22, 31
Pusino, I., 128 n.
Pythagoras, 81

Quintilian, 96

Rashi, 67
Reed, A. W., 60 n.
Regius, Urban, 139
Remigius, 134 n., 146
Renaudet, A., 129 n., 133 n., 142 n., 146 n., 147 n.
Reuchlin, Johann, 127 n., 137, 150; is a judge, 70; legal interpretation, 75; his teacher in Greek Hermonymus of Sparta, 93; on his vocation, 70; Bible studies occupy a central place in his life, 63

223

GENERAL INDEX

Reuchlin, Johann (cont.)
two aspects of his work, 71; the synthesis of these aspects, 82–4
his philological studies, 71–80; is a grammarian not a theologian, 78; his view on the relationship between philology and theology, 78–9; importance of the study of grammar, 82–3; demands freedom of criticism, 88–9; quotes canon law for the defence of his view, 72, 74, 75; publication of *De Rudimentis Hebraicis*, 66–7; *De Rudimentis Hebraicis*, 76–8; method of interpretation, 72–4, 81 n.; reasons for learning Hebrew, 71, 71 n., 78–80, 82; view on Jews teaching Hebrew in Germany, 65, 65 n.; on Jewish commentators of the Bible, 75, 77, 78, 80; Jewish philosophy is derived from God, 81–2; Hebrew is the holy language, 83, 84; only the Hebrew text of the Old Testament is correct, 72–4, 78; on the origin of the Hebrew vowel system, 72–4; scarcity of Hebrew books in Germany, 68, 68 n.; his knowledge of the Talmud, 80; his advice concerning the burning of the Talmud, 88
his cabbalistic studies, 80–4; the meaning of the Cabbala for Reuchlin, 80–4; the Cabbala is identical with Pythagoras's philosophy, 81; Cabbala is the source of philosophy, 81; his cabbalistic studies are criticized by humanists (Colet, Erasmus), 87, 114
humanistic defence of Reuchlin (*Epistolae Obscurorum Virorum*), 90; similar tendencies between Reuchlin and Erasmus, x, 78, 86, 88 n., 94, 126, 127, 127 n., 137–8, 150, 154, 210; Erasmus's attitude towards Reuchlin, 78, 78 n., 87 n.; opposed to Luther, 86; is thought to be favourable to Luther, 90; his work is used by Luther, 167, 177, 178, 178 n., 189; attitude towards the Schoolmen, 75–6, 85; his studies attacked by opponents (Pfefferkorn, Collin, Hochstraten, the Universities), 87–91
on translation, 71–2, 78, 85; Reuchlin as a translator, 84–6
attitude towards traditional exegesis and the Fathers of the Church, 76, 77–8, 79; attitude towards the Vulgate, 4, 71, 78–9, 84, 85; critical of Augustine, 77–8; on the Septuagint, 73

Revelation of St John, *see* Bible, New Testament, Revelation

Revised Version, *see* Bible, Editions and Translations, Complete Bibles

Rheims Bible, *see* Bible Editions and Translations, Complete Bibles

Rhenanus, Beatus, 93
Rhijn, M. van, 56 n.
Richard, P., 10 n., 159 n.
Richter, A. L., 10 n., 60 n., 159 n.
Rietschel, G., 13 n.
Riggenbach, B., 65 n.
Robinson, Wheeler H., 5 n.
Roerer, Georg R., 202 n.
Romans, Epistle to the, *see* Bible, New Testament, Epistle to the Romans
Roy, William, 158
Rückert, H., 186 n.
Rufinus, 36

Sallust, 96
Saul, 173
Schade, L., 29 n., 30 n.
Scheel, O., 168 n., 169 n., 172 n., 185 n., 189 n., 193 n., 195 n., 210 n.
Schelhorn, C., 78
Scheurl, Christoph S., 205
Schmidt, Ch., 104 n., 105 n.
Schubert, H. von, 167 n., 175 n., 186 n., 187 n., 188 n., 189 n., 190 n., 200 n.
Schulte, F., 10 n., 159 n.
Schwarz, Peter (Nigri), 63, 64, 66, 67 n., 68–9
Schwarz, W., 35 n., 93 n., 141 n., 178 n., 190 n., 208 n.
Scotus, *see* Duns Scotus
Scriptus, Paulus, 55
Scudder, J. W., 56 n.
Seeberg, R., 169 n., 172 n., 193 n.
Seebohm, F., 109 n., 115 n.
Septuagint, *see* Bible, Editions and Translations, Old Testament
Sidonius, 96
Simon, R., 163 n.
Sirlet, Cardinal, 145 n.
Smalley, B., 47 n., 69 n.
Smith, C. D., 10 n.
Soncino, family, 67
Sophocles, 141
Soury, J., 192 n.

GENERAL INDEX

Spalatin, 199, 205, 210
Spicq, P. C., 47 n.
Stapulensis, *see* Faber Stapulensis
Statius, 96
Steinbach, Wendelin, 52
Stokes, F. G., 87 n.
Stunica, *see* Zúñiga
Sunnia, 34, 34 n.
Symmachus, *see* Bible, Editions and Translations, Old Testament

Tauler, 189 n.
Terence, 96
Thackeray, H. St J., 17 n.
Theocritus, 92, 92 n.
Theodectes, 19
Theodotion, *see* Bible, Editions and Translations, Old Testament
Theophylact, 144 n.
Theopompos, 19
Thessalonians, *see* Bible, New Testament, Epistles to the Thessalonians
Thomas Aquinas, 9, 110, 112, 134, 146, 150, 151, 152, 152 n.
Tibullus, 96
Timothy, *see* Bible, New Testament, Epistles to Timothy
Trent, *see* Council of Trent
Tripoli (Oea), 38
Troeltsch, E., 183 n.
Tyndale, *see* Bible, Editions and Translations, Complete Bibles

Valla, Laurentius, 61, 72 n., 78 n., 91, 96, 97, 106, 116, 132, 132 n., 133–4, 136, 137, 144, 150, 152, 158
Vatican, *see* Council of the Vatican
Vergil, 96
Vergil, Polydore, 121 n.
Vetus Latina, *see* Bible, Editions and Translations, Complete Bibles

Vienne, *see* Council of Vienne
Vinsauf, *see* Geoffrey de Vino Salvo (Vinsauf)
Vitrier, Jehan, 109 n., 110, 128 n., 131 n.
Vogelsang, E., 172 n.
Vollmer, H., 13 n.
Voltaire, 114
Vulgarius, *see* Theophylact
Vulgate, *see* Bible, Editions and Translations, Complete Bibles

Wackernagel, R., 93 n., 142 n.
Walde, B., 67 n., 70 n.
Ward, Sir Adolphus, 70 n., 87 n.
Watson, Henry, 58 n.
Weigle, Luther A., 2 n.
Weisinger, H., 97
Weiss, N., 172 n.
Weiss, R., 93 n.
Wendland, P., 17 n., 22 n.
Wessel Gansfort, 56–7, 58
Westcott, B. F., 13 n.
Westminster, Archbishop of, 3, 9 n.
Wiegendrucke, Gesamtkatalog der, 47 n., 67 n.
Wilkins, D., 13 n.
Wille, C., 71 n.
Wimpfeling, Jacob, 97 n., 104, 104 n., 105, 105 n., 122, 123
Worsley, John, *see* Bible, Editions and Translations, New Testament
Wyclif, *see* Bible, Editions and Translations, Complete Bibles
Wyon, O., 183 n.

Zarncke, F., 58 n.
Zúñiga, Diego Lopez (Stunica), 164, 164 n.
Zwingli, Ulrich, 66

220.52   67063
S411p

Schwartz, Werner
Principles and problems of
  Biblical translation

220.52   67063
S411p

Schwartz, Werner
Principles and problems of
  Biblical translation

23 APR 69   2133
14 DEC 70   DEC 12 '70   6300
         SWITZER MARGUERITE
23099   SEP 9 '74
Name David Trotter
OCT 26 1988   3058   Box 30-L

Memorial Library
Mars Hill College
Mars Hill, N. C. 28754